A MODERN APPROACH IN GEOGRAPHY

About the Author

Dr. P.B. Saxena (*b. 1941*) took his Ph. D. in Geography from Kanpur University. His doctoral study was on "Dehra Den Valley: A Study of its Resources and Population". He has published a number of research papers in reputed journals. He is currently working on a project of the UGC on "Ecological and Integrated Area Development with Special Emphasis on Eradication of Poverty and Unemployment in the Garhwal Himalayas". His applied work in Geomorphology has been included in Concept,s Internationals Series in Geography (Perspectives in Geomorphology) Ed. By H.S. Sharma.

Dr. Saxena is at present a Senior Lecturer in Geography, D.B.S. College, Dehra Dun.

A MODERN APPROACH IN GEOGRAPHY

FOR THE EVALUATION OF SOILS AND LAND-FORM SYSTEMS OF LANDUSE PLANNING IN HIMALAYAN ECOSYSTEM OF THE ALAKNANDA BASIN

P.B. SAXENA

CONCEPT PUBLISHING COMPANY PVT. LTD.
NEW DELHI-110059

ISBN: 978-81-7022-012-1

First Published 1987

Published and Printed by

Concept
Since 1974

Concept Publishing Company Pvt. Ltd.
A/15-16, Commercial Block, Mohan Garden,
New Delhi-110059 (India)
T : +91-11-41101460
E : publishing@conceptpub.com, *W*: www.conceptpub.com

FOREWORD

It is a healthy trend that interdisciplinary approach is gaining ground in the field of Geography. The present work integrates the findings of Geomorphology, Pedology and Photo-interpretation. It is a planning-oriented work, supported with recent techniques like remote-sensing and fractional code method. It is hoped that this work will prove to be beneficial to the post-graduate students and research scholars in Geography, Geology, Ecology and Agricultural Sciences as it embodies the latest approach towards the functioning of the ecosystem without discarding much of the useful knowledge in Ecology gained during the last seventy-five years.

Dr. P.B. Saxena deserves our congratulations for having produced this splendid work on the Alaknanda Basin in Garhwal Himalaya.

Nitya Nand

Reader and Head
Department of Geography
D.B.S. College, Dehradun

18th Feb., 1985

ACKNOWLEDGEMENTS

I am deeply indebted to Dr. D.P. Saxena, M.A., Ph.D., Reader and Head of Geography Department, D.A.V. College, Kanpur under whose valued inspirations this project work was started.

I am thankful to Dr. Ram Mohan Singh, M.A., Ph.D., LL.B., Principal, D.B.S. College, Dehra Dun who gave all sorts of facilities for my lab. work in the college. He also encouraged me for publication work on Garhwal Himalaya.

I record my sincere thanks to Dr. Nitya Nand, M.A., Ph.D., Reader and Head, Geography Deptt., D.B.S. College, Dehra Dun for his valued discussions on the subject matter and the nature of work. He also gave me incentives for intensive field studies of the region. I express my gratefulness to Dr. G.S. Gosal, Prof. and Head of Geography Deptt., Punjab University, Chandigarh who approved my scheme of work and encouraged me to work on the lines of remote sensing techniques.

I am also highly grateful to Prof. H.S. Ayer, Soil Division, Prof. B.M. Singh, Soil Conservation Division, Prof. A.K. Roy, Geology Division and Prof. M.D. Sheda, Forestry Division, Indian Photo-Interpretation, Kalidas Road, Dehra Dun who extended their valued cooperation in the interpretation of varied aspects from satellite imagiries and photographs.

I feel very much grateful to my research scholar Shri R.K. Saxena, Sr. Soil Scientist, National Bureau of Soil Survey and Land Use Planning, Nagpur and Dr. J.P. Yadav, Asstt. Prof. in Soil Division, I.P.I., Dehra Dun who helped to analyse my soil samples. I also feel pleasure to express my thanks to Sri Satya Prakash, Naval Hydrography who helped me to get my maps prepared.

I very sincerely acknowledge with thanks the assistance and encouragement which I received from my colleagues and relatives. I am also very much thankful to my teacher research scholars Shri Satya Prakash and Prof. M.S. Anantharaman and

Prof. P. Bahukhandi who accompanied me in my field study traverses of the Alaknanda Basin.

In the end I take an opportunity of expressing my gratitude to the Director of ICSSR, New Delhi who, on the basis of quality of my published work awarded me this Pilot Project on the Alaknanda Basin, Garhwal Himalaya.

P.B. Saxena

CONTENTS

LIST OF FIGURES

LIST OF TABLES

ONE

INTRODUCTION

The Alaknanda Basin, which ranges from 300 m to 4800 m altitude comprising an area of 10880 km^2, consists of the eastern part of the Garhwal Himalaya. The drainage systems of the Alaknanda Basin due to being fed by glaciers have glacial, glaciofluvial and intermontane fluvial scenic landscapes. Out of the total area of the Basin, 433 km^2 is under glacial landscape and rest of 2888 km^2 under fluvial landscape. Thus the glacial fluvial agencies of the valley have enormously evolved diversified landscapes comprising unique features of river-confluences (Prayags) with agricultural terraces, incised meanders with interlocking spurs and unique glacial 'U' shaped valleys with retreating glaciers (confined to the receissional morrains). In addition to these, the Basin is well dressed with green vegetal cover, its head is crowned with ever shining snow and ice and its landscapic personality is ornamented with varied minerals and soils. The valley being well endowed with various land resources in plenty by way of heavy precipitation, soils, vegetation and minerals, is agriculturally undeveloped and geographically unexplored. The geonomic life of the region, which is primarily based on agriculture, is lagging behind because of the haphazard use of the agriculture land, lack of means of irrigation and the neglect of soil conservation measures.

The Alaknanda Basin consists of numerous resourceful valleys of its tributaries which are of great significance for agriculture. Various factors like relief, climate, soils and agronomic practices determine the present agricultural progress which is below the expectations in the region. Of the total land, about 40 per cent is arable land and rest of 60 per cent is under forest, glaciers and other non-agricultural uses. Out of the non-agricultural land 25 per cent land may easily be reclaimed for cultivation if means of irrigation, good management of the agronomic practices and soil conservation measures are provided. Of the total arable land, only 15 per cent land is irrigated and rest 85 per cent of the upper terraced and hill topped lands remain unirrigated due to lack of means of lift irrigation. In the valley the land use and cropping patterns still require slight ecological adjustment, with the result the yield per acre decreases and the bare fallow land goes under severe erosion. The poor Katil land is always left as bare fallow for 3 to 4 years due to lack of means of lift irrigation and it is found to be gullied and revined in general course. About seven cycles of crop rotation prevail freely in the valley consuming long course of period from 2 to 9 years.

It is generally observed that the low capable land of the rainfed areas of the valley is being deficient in both organic and inorganic properties. The organic properties (humous) are exhausted by using land without any special practices and inorganic properties are depleted by it leaving the land barefallow after ploughing two or three times for maize and rice. In fact the efficiency of soil is being reduced by haphazard use of land, lack of means of irrigation and the neglect of soil conservation measures.

In brief the study reveals the geographical appraisal of the soil resources in relation to the landuse practices in the Alaknanda Basin. The soils of the Alaknanda Basin are generally gravelly and mostly acidic with pH values ranging from 6.0 to 7.0. However, in many localities the soils, developed from the limestone and gypsum rocks are very fertile and suitable for cultivation of hill rice and wheat. Podzolic soils of the coniferous zone of the valley are leached and poor in fertility, their pH values vary from 6.0 to 6.6. The soils of the Alpine zone of the Alaknanda valley are generally stony gravelly and acidic with low pH value i.e., below 6.0 because the loose boulders of the Blanis series are embeded with silt in the lateral structures of the valley. Consequently the rate of infiltration remains higher varying from 5.87 cm, 3.83 cm, 3.82 cm for 1st, 2nd and 3rd hours respectively under normal rains of 1423 mm per annum. Overall the soil cover of the valley is shallow and stretched in patches with thickness ranging from 0.1 m to 0.5. The percentage of the kankar in the soils varies from 20 to 45 and the amount of the clay varies between 8.3 and 18% descending from the upper terrace to the valley bottom.

On account of shallow cover of the soils with loose boulders and gravels, under heavy rains the valley looses an average amount of 52.75 tons of soil per acre per hour.

AIMS AND OBJECTIVES

The proposed study aims to appraise the soils of the valley by analysing their morphology, taxonomy and capability after collecting samples from varied agricultural lands and delimit the basin into land capability regions to highlight their impact on present land use system.

The further objective of the study is to evaluate the agricultural land use efficiency and finally regionalise the valley into landuse efficiency zones to diagnose the geocatetrophic factors pinpointing their intensity of agricultural resource depletions and treat them with the negotiations of the ecological measures by making due adjustments in cropping pattern and adopting suitable types of horticulture and apiculture. Preventive measures of soils conservation are basically needed to achieve maximum and multiple uses of the land and soil through green, white and blue revolutions. Ecologically the land and soils of the region need a rational planning for agricultural, horticultural and agro-industrial developments. Suitable sites for lift irrigation scheme (hydrum) shall also be studied to irrigate the uplands for increasing land pruductivity in the region. The study of the practices of legumes, kodju (long creeper) grass, Agave sislana (Ram bans) on scarp ridges, bunds of terrace and hill top shall also be made, to prevent the soil erosion and protect the crops from the wild life.

METHODOLOGY

In the present work, the author has applied recent research techniques of remote sensing and quantification to justify the relationship among climate, physiography, soils and landuse patterns in the Alaknanda Valley (Garhwal Himalaya). In the beginning a detailed reconnaissance survey of the Alaknanda Basin (extending from 30° N to 34° N lat. and 79° E to 79° 49'E long with the area of 10880 km^2) was operated on the scale (R.F.I. : 50,000) to depict the physiography, parental rock of soils and broad land use patterns. Field study traverses were also conducted along the main tributaries of the Alaknanda to confirm the results as worked out from the topographical sheets and land-sat imagiries of the region. After completing the reconnaissance survey, network of the landform systems, land use systems and drainage systems were broadly deleneated on large scale maps. The important physiographic elements (relief, slope and aspect) which directly govern soils were analysed from maps and land-sat imagiries and they were randomly checked by field study traverses. Due weightage was given to these elements to workout the physiographic index number on the following method:

$$\text{Physiographic Index No.} = \frac{(\text{Relief} + \text{Slope}) \times \text{Aspect}}{\text{Soils}}$$

With the help of the physiographic index number, the relief was categorized to appraise the influence of relief units over soil formation and existing land use of the region. Similarly the other parameters like climate, land use and varied soil properties were duly categorized and given them due weightage to prepare a legend of the resultant map as given below:

$$\text{Legend} = \frac{\text{Relief + Climate + Slope + Aspect + Land Use}}{\text{Soils}}$$

The above work was supported with the collection of the primary data on agricultural system in the valley through questionnaires and schedules which were confirmed by personal visits, investigations and interviews. In such a rugged terrain like Himalaya, field investigations are time consuming and arduous, therefore aerial photographs and land-sat imagiries were also consulted as bases for collecting reliable data. To measure the dip, strike distance, gradient, relative humidity and altitude of relief sub-units which are responsible for the nature of soil genesis, pattern of land use, field instruments like clinometer compass, pedometer, brunten compass, hygrometer, altimeter and soils colour chart were used in field studies.

In the present survey, the data related to climate and hydrology have been collected from various sources, whereas soil and land use systems were actually surveyed in the field. Various soil samples, studying their profiles of soil horizons were also collected from representative physiographic units. Soil, due to being a three dimentional body and as such it may not be recognized on the aerial photographs yet physiographic analysis may provide sufficient clues for delineating boundaries among unlike soils.

Systematic photo interpretation for soil surveys, following physiographic approach, differentiate the physiographic units, lithology, aspect, altitude and existing land use. Each unit and its sub-units were tabulated.

The soils of the surveyed landscapes were classified according to the comprehensive system of soil classification U.S.D.A. (Soil Taxonomy 1975) upto family level and their association. Soils were identified at the highest category level as mentioned below:

(i) Entisols, (ii) Inceptisols, (iii) Mollisols, (iv) Alfisols

In addition to the above studies of soils, some more surveys were conducted to separate the associates and to identify phases like soil texture, stoneyness, soil-depth, soil-profiles, soil-slopes, organic-matter and soil-erodibility. All these parameters of the soil survey, were analysed and due weightage was given to them to determine the land evaluation factor No. which is essential for land capability classification to strengthen the land use classification and its rational planning. It is mentioned below

$$\text{Land Evaluation Factor No.} = \frac{\text{Soil Index No.} \times \text{Physiography index No.}}{100}$$

Thus the author has produced a systematic land use table for the mountainous terrain in which all parameters related with soil, physiography aspects and climate were given properly. To make the detailed study of the agricultural land use, the case study of the lower part of the Mandakini valley was made. The case study of the region was supported with the 'Unit Area Method'[1], of land classification for agricultural land capability. The 'Unit Area Method' is an adaptation of 'Fraction Code Method' i.e., the representation of qualities of the land significance in digits in which fractional code notations are applied to units. The physical units (slope, relief, aspect, land system) are allotted by digits in the denominator of the fraction and soil factors responsible for agricultural capability (soil texture, depth, organic matter, erosivity inherent fertility) were noted by digits in numerator of the fraction. After analysing the results of the unit aea method, the degree of capability of land was appraised to classify the land for agricultural planning. In general, the region has been categorised in six land classes.

To rationalise the agricultural land use planning of the region, land depletion factors were also studied from remote sensing techniques and they were confirmed through field studies. The author has also analysed the intensity and distribution of land depletion (caused by landslides, rock slips, mudflows and soil creeps) and then suggested preventive measures to cure landforms with ecological negotiations.

Reference

1. The Unit Area Method of land classification for agricultural land capability as developed by Hudson and collaborators used in the land classification section of T.V.A. has been modified by the author.

TWO

PHYSICAL LANDSCAPE AND PHOTO-INTERPRETATION OF LANDFORM SYSTEMS AND ELEMENTS

GEOIDENTICAL CONDITIONS

The Alaknanda valley comprising of 10880 km^2 area of the Garhwal Himalaya extends from 30°N to 30°47'N latitude and 79°E to 79°49'E longitude. The valley is surrounded by the sky kissing peaks of the Garhwal Himalaya ranging from 3000 meter to 4800 meter and above. The nature has endowed the valley with glacial, fluvioglacial and fluvial landforms, green vegetation, soils and minerals, even the rural life of the valley primarily based on agriculture, is still deplorable. Agricultural development of the valley generally lags behind on account of severe resource depletion, which occurs due to multiplying results of resource hazards like excessive soil erosion, mass wasting, soil creeping, torrential runoff, slope failure, severe infilteration as a consequence of deforestation, unscientific quarrying over blasting, overgrazing and over fallowing. These prevailing unrationalised factors cause major landslides.

Geologically the Alaknanda valley consists of three tectonical separable major litho-stratigraphical units know as the Dudhatoli Group, the Garhwal Group and the Central Crystalline group. The zone between Deoprayag and Koteshwar constituting of moderately metamorphased rocks particularly phyllite and quartzites forms the Dudhatoli Group, which is separated by the North Almora Thrust trending almost in NW-SE direction. Further the rocks of the Garhwal Group are exposed upto Vishnu Prayag and are primarily quartzites, slates, schists and carbonate rocks occasionally intruded by metavolcanics. The Garhwal Group which is separated by the Main Central Thrust existing in NW-SE direction, consists of several shear and fracture zones viz., Kaliasor Nand Prayag and Belakuchi, which have highest frequency of land slides. The Northern Zone existing from Vishnuprayag to the village Mana and its environs along the river Alaknanda and its tributary Dhauliganga consists of schist, gneiss and granite of the Central Crystalline Group

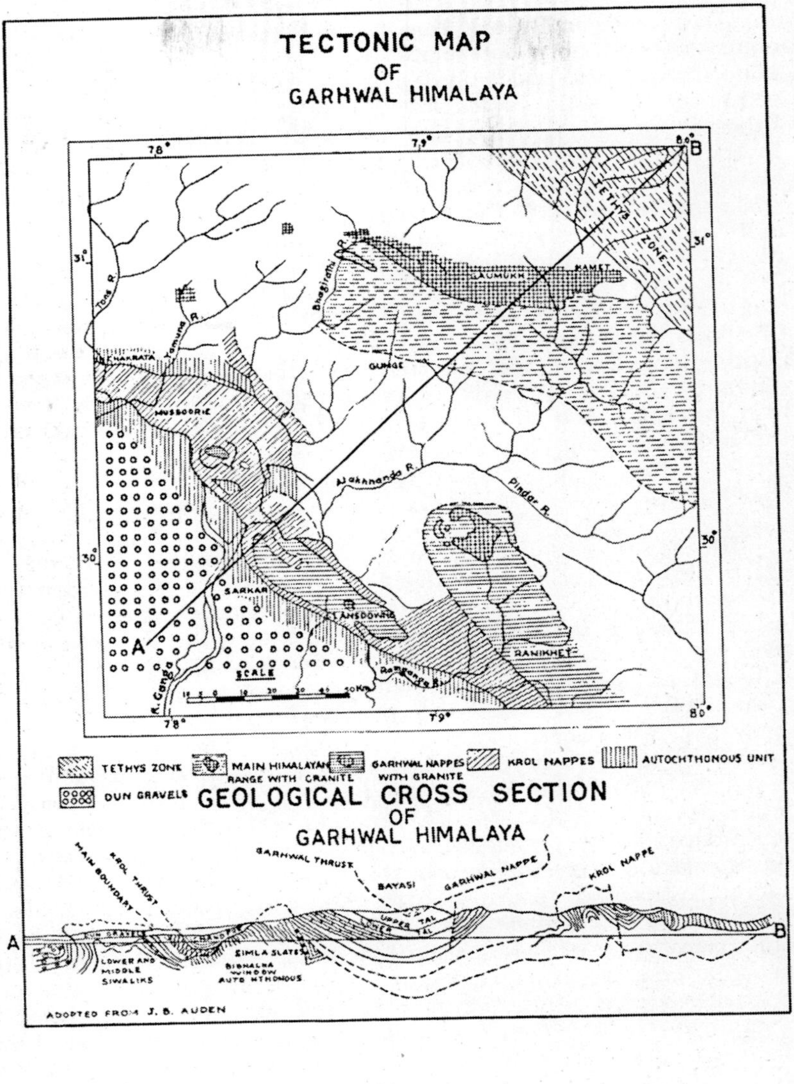

Fig. 1: Tectonic Map of Garhwal Himalaya

which rests over Garhwal Group due to Central Crystalline Thrust.

Thus tectonically soil formation processes (serecitization and chloritization) were estimulated by thrusting, faulting and crumpling movements under humid conditions of the valley. Consequently the soils varies in structure, colour and depth in all lithostratigraphical units. The mollisols (deep soils) are observed in the valley terraces of the villages of Augustmuni, Gauchar, Maletha and Nagrasu etc. (below 1000 metre m.s.l.) in between Alaknanda thrust and the North Almora thrust. The deep glacial drift soils are generally spread in patches (ranging from 3000 to 4800 metre high m.s.l.) as observed in the villages of Dhauliganga under Central Crystalline Group bordered with the Main Central Thrust and Dhar-Martoili Thrust. The incepte sols (Shallow soils) are developed on the southern aspect of the concavoconvex slopes of the low mountains and aute-sols (Shallow soils) are found on moderate to steep slopes of the high mountains due to least effect of the tectonic movements in the region.

Drainage

The river Alaknanda, which rises from the snout (6615 metre high from m.s.l.) of the twin glaciers of the Bhagirath-Kharak and Satopanth, flows across the ranges of Greater and lesser Himalaya to form its last confluence with the Bhagirathi river at Deoprayag (725 m high from m.s.l.). The valley has achieved all stages of youth with maximum river order system from first to six resulting in five prominent confluences (prayags). The regression trend of the drainage orders indicate that the fifth and sixth order systems are not linearly regressed with lower order showing immaturity of river drainage net work. The Alaknanda drainage net work is embroidered with prominent subsequents like the Dhauliganga, the Nandakini, the Pindar and the Mandakini rising from Kamet glacier, Nandakot (snowpeak) Pindari and Chaurabari glaciers respectively. The valley due to being subjected to various tectonic movements of asymmetrical folding, reverse faulting and thrusting consists of multivariate structure of drainage patterns. In the crumpled, thrusted and faulted strata of the Central Crystalline Thrust zone, the valley comprises of perennial flow of water in the form of contorted drainage. Generally tributaries of the Alaknanda river following the direction of thrusts (Main Central Thrust, Alaknanda thrust and North Almora Thrust) trending from NW to SE crossing the river Alaknanda nearly in rectangles, contributed their waters at rectangular and subrectangular confluences (prayags) to evolve subparallel drainage patterns. The valley also comprisies of prominent imprints of net work of underground drainage of the Patalganga. The dendritic net work which is generally evolved by the insequents of the head stream exists near the sources of the tributaries of the Alaknanda.

Thus the multivariate drainage net work of the river Alaknanda has sculptured and engraved the valley in very complicated landform like mountains (High mountains, Low mountains, Very Low mountains) and river valleys (incised meanders, river benches, flood patches and escarpments etc.). The main subsequents of the Alaknanda river forming subparallel drainage net work have transversely dissected and gullied highhills and hill side slopes. But the engrown meanders as interpreted from the land-sat imagiries and photographs in this net work (below the elevation of 1000 m from m.s.l.) have broad paired terraces of mollisols (Deep soils), rock benches and flood patches in the Lower Alaknanda valley. The underground drainage net work activates the land depletion through under-cutting action of river water.

The glaciated areas which have been sculptured wholly or partly by the Bhagirathi-Kharak and Satopanth, Kamet, Pindari and Chaurabari glaciers, comprising of 25.13 per cent land of the Alaknanda catchment, contribute to a great extent as an intermittent feeders to the river Alaknanda and its main tributaries to give them antecedent status.

Climate

Climatic conditions for rugged mountainous terrain of the Alaknanda valley, differ from place to place due to the change of relief. A degree of diversity in relief contributes to a great extent to the irregularities of climatic elements which significantly influence the land use patterns. In the mountainous terrain, temperature plays an important role to determine the limit of crops rather than rainfall.

The amount of mean-annual rainfall reduces from 2000 mm to 250 mm with the rising trend of altitude varying from 1000 to 4000 m and above the elevation of 4000 m the snowfall occurs partly to wholly. Similarly the mean annual temperature decreases with the increase of altitude as recorded 20°C to 12°C in the lower reach of the Alaknanda basin ranging from 725 m to 1200 m and it further decreases from 15°C to 10°C in the middle reach of the Alaknanda basin ranging from 1200 m to 2400 m altitude from the sea level. In the upper reach of the Alaknanda basin the temperature reduces from 2400 m to 5000 m and above. (Table V).

Thus the abrupt change in the climatic elements in the valley, influences the soil genesis land use patterns. In such a mountainous terrain, the temperature plays a very important role to delimit the crops zones than that of precipitation, because temperature remains not available in sufficient amount in the higher altitudes to ripen the crops with the result the crop harvesting period differs altitudinally with the change of temperature (the amount of sunshine) e.g., the harvesting period of wheat varies from May to August in the Alaknanda valley according to the rise in the altitude from about 1000 m to 2700 m and above.

Thus the valley consists of mainly three distinct climatic zones: (I) The subtropical climatic zone exists below the elevation 1200 m (from sea level) where soil moisture regime is usually ustic having thermic temperature class C_{22}, (II) Cool temperate climatic zone ranges from the elevation of 1200 to 2400 m where soils have udic moisture regime and thermic to mesic temperature class (C_3) and (III) Cold zone - generally occurs above the elevation of 2400 m where soil moisture regime is aridic to ustic and have frigid temperature class (C_4) upto the elevation of 2000 m. Above the elevation of 5000 m the soils have usually aridic soil moisture regime and frigid to cryic temperature class (C_5) where snow remain permanently. (Table V).

INTERPRETATION OF LANDFORM SYSTEMS AND THEIR ELEMENTS

Generally the pattern of landform system signifies the type, mode, and behaviour of morpho-units of the physique of the region. The identification of morpho-units in the Himalayan terrain is very important to analyse the parental material of soils and to rationalise land use with its potentiality.

Physiographically the Alaknanda valley consists of two major landform systems as subdivided below:

1. Landform system

1.1. Glaciated Landscape.
1.2. Fluvial Landscape.

1.1. Glaciated landscape occupies an area of 2504.99 km^2 which consists of 23.13 per cent land of the Alaknanda catchment. The glaciated landscape which has been sculptured by the glacial unctions, ranges from 4800 m to 8000 m elevation from sea level.

1.2. Fluvial landscape consists of mountain and river valley structures covering an area of 8375.17 km^2, i.e. 76.87 per cent of total area of the region. The fluvial landscape has been further subdivided after considering the nature of physiographic elements of relief, slope and aspect to evaluate distinctly soil and land uses in the valley.

1.2.1. Mountains

The total area of the mountains ranging from 1000 m to 4000 m and above from sea level consists of 7649.19 km^2, i.e., 82.22 per cent of the total area of the Alaknanda basin. The mountains have been further subdivided into the following classes:

		Range of altitude (m)	Code No.	Weightage
1.2.1.1	Very low Mountains	1000	1	20
1.2.1.2	Low Mountains	1000-2000	2	15
1.2.1.3	High Mountains	2000-3000	3	10
1.2.1.4	Very High Mountains	3000-4000	4	05
1.2.1.5	Extremely high Mountains	4000	5	01

1.2.2. River valley structures

The Alaknanda and its main tributaries consist of very complicated structures of their valleys as characterised by the incision process of the rivers. The river valley structure has been further subdivided. The area under river valley structures comprises of 1433.15 km^2 of the Alaknanda catchment, i.e., 17.02 per cent of the total area of the basin.

		Code No.	Weightage
1.2.2.1.	River terraces: 'T'	1	30
1.2.2.1.1.	Structural terraces - River benches (made on spurs) T_2	2	35
1.2.2.1.2.	Alluvial terraces Repose hillslopes (Flood patches)T_3	3	22

1.2.2.2. Talus cones (C) consist of conical forms of unconsolidated debris. The lower reaches of talus cones are usually cultivated and upper reaches are dry and mostly found under scrubs and bushes. The total area under fully matured talus cones, consist of 48.64 km^2 of the Alaknanda basin.

Physiographic Elements

Soils and land use are not only influenced by macro/micro climatic conditions but also physiographic elements of relief, slope and aspect. The due weightage to the cadre of elevation has been given while both the factors relief and slope have been taken as combined factors and aspect as a modifier factor.

Relief treats a comprehensive structure of the physical features of the region relatively high and low and from the river base level.

Relief has been categorised in the following classes :

S. No.	Relief	Properties	Weightage
1.	High Relief	High hill with convex slope (Waxing slope)	50
2.	Normal Relief	Hillside with straight slope (Free face slope)	30
3.	Low Relief	Foothills with conve slope (wanning slope)	20

Slope: Slope is an important factor which controls runoff and removal of sediments and weathered debris. In the present studies the slope has been classified in the following categories and due weightage has been given to them.

Code No.	Slope categories	Unit (%)	Weightage
1.	Flat to gently sloping land	7°	40
2.	Gentle sloping to moderate	7° - 15°	30
3.	Moderate to steep	15° - 30°	20
4.	Steep to very steep	30°	10

Aspect: The soil moisture and temperature are generally affected by the aspect factor. Since a major part of the study region occurs under forest management. The northern aspect, identified for forests whereas southern aspects suitable for cultivation and grass lands have been represented by fractional code as given below:-

		Weightage	
1	Northern Aspect	0.50	(North-Northeast-East)
2	Southern Aspect	0.25	(South-Southwest-West)

THREE

SOIL MAPPING
Its Genesis and Classification

Normally soil map provides mapping units of association of soil families/series. Soil mepping units are based on characteristics of geology, physiography and climate. Tor land capability classification, soil mapping reconnaissance survey plays a very important role. In the hilly terrain of the Alaknanda valley, where soil formation is a rapid progress due to its damp climate and dense vegetal cover. Soil survey needs prior attention before evaluating its rugged terrain. The soil formation of the valley varies with the change of its lithology and tectonic features like faults thrusts and crumples. The lithological structures of the valley is influenced by the parallel major thrusts trending from North-west to South-east like North Almora Thrust, Alaknanda Thrusts, Main Central Thrust and Dhar-Martoli Thrust. In the Alaknanda valley which consists of gneiss, granite phyllite schist and quartzites, sericitization and chloritization processes transform them into loamy sand, micaceous, dark grey clayey, sandy and stoney soil respectively. In addition to these the patches of shales and metadolerites are weathered into clayey and acidic soils.

SOIL GENESIS

The nature and characteristics of the soil at different locations and different situations are mostly developed as a result of interactions of soil genetic factors: climate, relief, vegetation, parental rock and time in addition to human influence. The evolution of parent landscape in the area under study is due to Himalayan orogeny and it is entirely hilly terrain. Thus the valley structures generally consist of both alluvial and diluvial materials. In the valley where broad terraces are formed, sediments have been deposited at some extent, therefore parental material is less important to form valley side alluvial terraces; but in the formation of structural terraces the process of mineral transformation pedo-chemical reactions and rock disintegration play an important role. The zone under investigation due to being located in the environs of the Main Central Thrust, the Alaknanda thrust and the North Almora Thrust still now undergoes mechanical weathering.

Out of fine soil genetic factors climate and topography as reflected in the form of slope, aspect and lithology have played a major role to form different type of soils. The vegetation plays a dominant role in the form of pedo-chemical weathering. The influence of climate accelerates the rate of chemical and mechanical soil genetic processes like sericitization, chloritization, kaolinization, pedo-chemical reaction and rock disintegration. The relief directly influences soil formation through its effects on drainage and runoff. The soil profile formations depend mainly upon the average amount of moisture in the soil. With the result, the soil on the steeper side have thinner solum and less distinct horizons, on the other side (gently sloping) concave slope have a thick solum and distinct horizons (Fig. 7). Thus alluvial soil is richly deposited on the valley side to form lower alluvial terraces in patches in the Lower Alaknanda Basin. The patches of deep soils which consist of structural terraces located on spurs are result of rock which undergoes chemical and mechanical weathering.

In the zone under study, it is worked out that chlorite schists dominantly comprise chlorite and biotite minerals which are more susceptible to the solution and hydraulic actions of water. The most effective mechanization is the addition of hydrogen ions to the rock which cause replacement of cations. The chemical weathering by mineral transformation (Sericitization, chloritization and kaolinization) mostly occurs on account of the type of lithological structure. Pedo-chemical weathering which occurs due to the nature of vegetation plays a dominant role in the area under study. Under sericitization process the chlorite mineral is converted into sericite which transforms finally into soils. The chloritization process converts the biotite mineral into chlorite and chlorite is further transformed into sericite (by sericitization) as mentioned above. Similarly the kaolinization process transforms the felspar mineral into clay but this process functions where granite gneisses occur. Chemical reactions are given below:

Sericitization

(1) $2\ Al(Mg,Fe)_5\ Alsi_3o_{10}(OH)_8 + 5Al^{3+} + 3si(OH)_4 + 3K^+ + 2H^+$

$3K\ Al_2\ Si_3o_{10}(OH)_2 + 10(Fe, Mg)^{2+} + 12H_2o.$

Chloritization

(2) $2K(Mg, Fe)_3Alsi_3o_{10}(OH)_2 + 4H^+ \ Al(Mg, Fe)_5Alsi_3)_{10}(OH)_{8+}$

$(Mg, Fe)^{2+} + 3SiO_2$

The geological succession of rocks around the village of Gauchar, Nagrasu and Bamoth support the pedo-chemical and mineral transformation processes of soil genesis as discussed below:

(i) Alluvial
(ii) Metabasics
(iii) Quartzite with thin inter calation of phyllite slate
(iv) Sericite - quartzite schist (withouinter clation of chlorite schist and basic silts).

Sericite quartz schist as available in zone are found dark green where they come in the contact with the basic rocks. Small patches of chlorite schists are observed with sericitic quartz and they are more foliated in between Srikot and Pansi.

Quartzites are fine to medium grained white to greish white in colour. These hard and compact rocks have been found converted into sistose quartzite near their contact with sericite quartz schist to the north of Bamoth and Gauchar.

The metabasics are fine grained dark green coloured rocks with distinct foliation planes. They are schistose in nature wherever schistosity is not developed but they appear very massive and forming narrow valleys.

Similarly the geological succession of rock around the village Augustmuni supports the above mentioned chemical processes as detailed below:

(i) Granite gneiss
(ii) Metabasics
(iii) Chlorite schist.

Biotite gneiss which is grey to greyish and black in coloured hard and massive rock, consists of chlorite and schist bands which are transformed into soils under sericitization and chloritization. The chlorite schists found near Augustmuni is completely sericitized. Metabasics which generally occur in chlorite schist and biotite gneiss, are light, dark green and massive and compact in nature as observed near Silli Chatti and Rampur. In the metabasics the band of biotite is transformed into chlorite by chloritization process and real metabasics remain very compact and massive.

The granite gneisses are greyish white and pinkish in colour and medium to coarse grained. Under Kaolinization process the felspar mineral of granite is converted into clay but its quartz is least affected. Hence the river becomes narrow towards north of Augustmuni village wherever granite and gneiss structure occur.

The factors of the soil gneisis in the area under study have distinctly influenced the phases of soil samples collected from representative physiographic sub-units as discussed below:

1. Soil texture

(i) Sandy
(ii) Coarse loamy
(iii) Fine loamy
(iv) Clayey loamy

2. Rockyness

Class - 0
Class - 1
Class - 2

3. Soil depth

Shallow - Paralethic/lithic contact within	- 25 cm
Moderate - Lithic/paralithic contact	- 25-50 cm
Deep - Lithic/paralithic contact	- 50-100 cm
Very deep - Lithic/paralithic contact	- 100 cm

4. Organic matter

High - More than 1 per cent organic matter in the top 25 cm (Mollic epipedon)

Low - Having less than 1 per cent organic matter in the top 25 cm.

5. Soil slope

(i) Gentle - limit
0-3 per cent Upper limit
5-18 per cent Lower limit

The rating of each soil factor has also been done:

(1) Soil texture: (Weightage 5-25)

		Weightage
(i)	Sandy skeletal	5
(ii)	Sandy loam	10
(iii)	Loamy skeletal	15
(iv)	Coarse loamy	20
(v)	Fine loamy	25

(2) Rockyness/stoneyness: (Weightage 10-25)

(i)	Class 0 (Nil to rockout crop) less than 20 per cent area in top soil	10
(ii)	Class I (Rocky soil) - 2 per cent to 25 per cent are covered by stones	15
(iii)	Class II (Very rocky) - 25 per cent to 90 per cent are covered by stones	20
(iv)	Class III (Rocky out crops) - above 90 per cent	25

(3) Soil depth: (Weightage 10-20)

	Classes	Weightage
(i)	Shallow paralithic/lithic contact within 25 cm	10
(ii)	Moderate lithin/paralithic contact within 15-50 cm	15
(iii)	Deep lithic/paralithic contact (within 50-100 cm)	20

(4) Organic matters: (Weightage 5-15)

	Classes	Weightage
(i)	High - more than 1 per cent organic matter in the top - 25 cm	15
(ii)	Low-having less than 1 per	

	cent organic matter in the tope - 25 cm	5

(5) Inherent fertility: (Weightage 10-25)

	Categories	Weightage
(i)	High base saturation	25
(ii)	Medium saturation	20
(iii)	Low saturation	15
(iv)	Poor saturation	10

(6) Erosivity: (Weightage 20-50)

	Categories	Weightage
(i)	Slight erosion	50
(ii)	Moderate erosion: Moderate loss of top soil generally by shallow channels and gullies	40
(iii)	Severe erosion: Severe erosion, severe loss of top soil by deep channels or gullies	30
(iv)	Very severe erosion: complete truncation of the soil profile and exposure of 'b' horizon	20

Besides the above, the soils samples collected from representative major land system units have been chemically and mechanically analysed for as noted in Table I.

From Table 1 it is clear that villages located on the lower terraces of the river valleys have thick cover of soils with higher percentage of organic matter and nitrogen, with the result the soil is fertile. For instance, the villages of Kaliasaur, Nagrasu and Gauchar which are located on the lower terraces in the Lower Alaknanda Basin, have broad alluvial terraces with considerable thickness and greater amount of organic matter i.e. (11.26%, 9.51%, 10.97%, nitrogen i.e. 0.56%, 0.47%. 0.53%) and pH value i.e. (7.00, 6.20, 8.00) respectively. As regards the agricultural land use pattern, it also differs according to the physiography and the soil capability. The soils of the said villages have good productivity of wheat, barley, gram, rice, Jhangora and marsa under two crops rotation a year in the valley.

Table I

Chemical Analysis of Soil Sample of the Alakhnanda Basin

Situs Structure of localities	Elevation in meter	pH value ue	Ec mm/ hr. cm	Organic matter	Nitrogen (%)	Lithostratigraphic groups
1. Mana village Glacial drifts	3000	6.50	0.10	2.40	0.01	
2. Malari glacial drifts	2760	5.40	0.11	2.52	0.3	Central crystalline group
3. Jelum structural terrace besides the spur	2700	7.05	0.05	2.60	0.13	
4. Joshimath structural terrace	1880	7.05	0.10	4.42	0.22	
5. Helang beside the repose sclope on talus	1800	6.50	0.11	5.00	0.30	Main Central Thrust
6. Pipalkoti structural terrace	1210	7.50	0.10	1.17	0.06	
7. Gauchar alluvial terrace (Flood plain)	1135	8.00	0.14	10.67	0.56	Garhwal Group
8. Nagrasu alluvial terrace	1000	7.20	0.13	9.51	0.47	
9. Kaliasaur alluvial terrace	660	7.00	0.11	11.26	0.56	Pudhatoh group

The soils of those villages which are located on the spurs and upper terraces (unirrigated have less amount of organic matter, nitrogen, with the result the soil becomes of poor quality. For instance, the villages of Barhgarh and Jelam, which are located in the Upper Alaknanda Basin, have very thin cover of soils (except in Malari village which stands on glaciofluvial drifts), low amount of organic matter i.e. (1.17%, 2.60%, 2.52%), nitrogen i.e. (0.06%, 0.13%, 0.02%) and pH values i.e. (5.70, 7.05, 5.40) respectively.

The said results of the analysis, indicate that the agricultural land of the Upper Alaknanda Basin is of poor quality and the land of the Lower Alaknanda Basin is of good quality. The soils of the villages of Belakuchi, Joshimath and Lata which are located on the river cut terraces in the Middle Alaknanda Basin, have less amount of organic matter i.e. 5.72%, 4.42%, 7.68% and nitrogen i.e. 0.22%, 0.38%. 0.06% with the pH values from 7.05 to 8.00.

CLASSIFICATION OF THE SOILS

The soils of the region have been classified according to the comprehensive system of soil classification U.S.D.A. Soil Texonomy - 1975) upto family level and their associations. The soils of the region fall under four orders: (i) Entisols, (ii) Inceptisols, (iii) Mollisols and (iv) Alfisols.

The classification of lower categorical level is schematically shown below:

(I) Entisols

Aguents
Psammaquents
Typic psammaquents
Coarse sa d mixed, Hyperthermic family

Orthents

Udorthents
Typic udorthents
Loamy, skeletal, mixed, hyperthermic, family
Coarse sandy, mixed, hyperthermic family
Lithic udorthents
Loamy, skeletal, mixed, hyperthermic family

Psamments

Udipsamments
Typic udic psamments
Mixed, hyperthermic family

(II) Inceptisols

Orchrepts
Ustochrepts
Udic ustochrepts
Fine loamy, mixed, hyperthermic family
Loamy, skeletal, mixed, hyperthermic family

Lithic Ustochrepts
Loamy, skeletal, mixed, hyperthermic family

Fine loamy, mixed, hyperthermic family

Eutochrepts

Typic eutrochrepts
Fine loamy, mixed, hyperthermic family
Loamy, skeletal, mixed hyperthermic family

Dystric Eutrochrepts

Loamy, skeletal, mixed, hyperthermic family
Coarse loamy, mixed, hyperthermic family
Fine loamy, mixed hyperthermic family

(III) Mollisols

Udolls

Hapludolls
Typic hapludolls
Loamy, skeletal, mixed, hyper-thermic family
fine loamy, mixed, hyperthermic family
Coarse loamy, mixed, hyperthermic family

Argiudolls

Typic argiudolls
Loamy, skeletal, mixed, hyperthermic family
Fine loamy, mixed, hyperthermic family

Ustolls

Haplustolls
Udic haplustolls
Loamy, skeletal, mixed, hyperthermic family

Argiustolls

Udic argiustolls
Loamy, skeletal, mixed, hyperthermic family
Fine loamy, mixed, hyperthermic family

Lithic Argiustolls

Loamy, skeletal mixed, hyperthermic family
Fine loamy, mixed, hyperthermic family

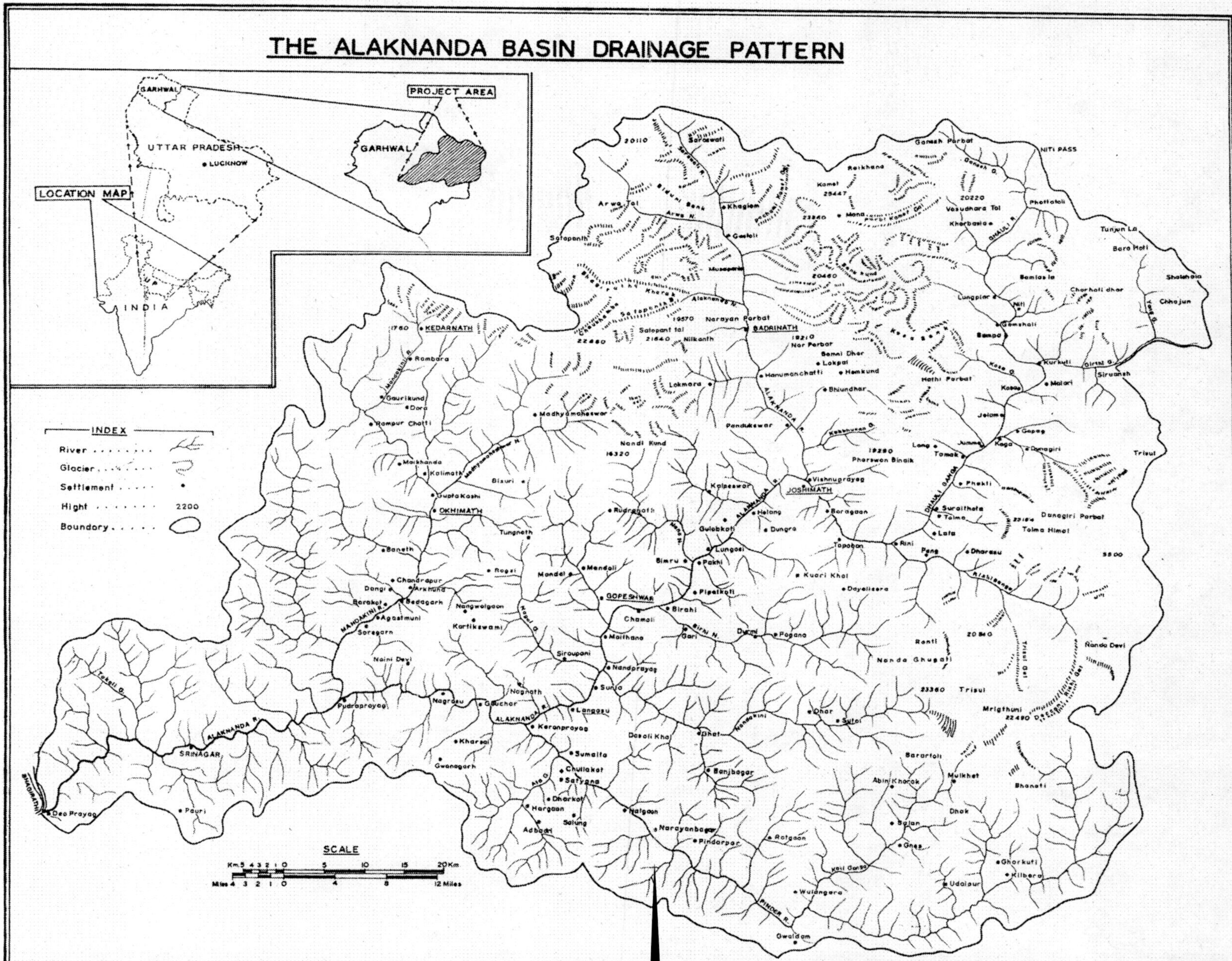

Fig. 3: The Alaknanda Basin Drainage Pattern Insets: Keymap, Delimitation of Sereal and Geographical Region

(Facing page 22)

(IV) Alfisols

Udalfs

Hypludalfs
Typic hapludalfs
Loamy, skeletal, mixed, hyperthermic family
Fine loamy, mixed, hyperthermic family

Mollic hapludelfs

Loamy, skeletal, mixed, hyperhermic family

Lithic hapludalfs

Fine loamy, mixed, hyperthermic family

Paleudalfs

Typic paleudalfs
Loamy, skeletal, mixed, hyperthermic family

Tropudalfs

Typic tropudalfs
Fine loamy, mixed, isothermic family
Loamy skeletal, mixed, isothermic family

Lithic tropudalfs

Fine loamy, mixed, isothermic family
Loamy, skeletal, mixed, isothermic family

Ustalfs

Haplustalfs
Udic haplustalfs

FOUR

LAND USE PATTERN AND AGRICULTURAL SYSTEM

The existing land use of the hilly terrain of the Alaknanda Basin in addition to climate primarily follows the nature of physiographic elements of relief, slope and aspect. Physiographically the region consists of mainly two types of landscapes: Glacial and fluvial. The fluvial landscape which has been studied in detail in relation with land use mapping consists of varied morpho-units of mountains (low, high, very high and extremely high mountains) and river valley (river benches, talus cones, escarpments and recent alluvial deposits). Thus the region due to being dissected and uneven finds abrupt changes in land uses.

LAND USE PATTERN

Of the total area of the Alaknanda Basin (10880.16 km^2) 86.7 per cent (9128 km^2) area occurs in Chamoli district, 6.8 per cent (797 km^2) in Pauri district, 6 per cent (699 km^2) in Tehri district and 0.5 per cent in Pithoragarh and Almora districts i.e., 50 km^2. Thus the valley is so large that varied land use patterns have been marked out (Table II).

Cultivated Land

In the catchment area of the Alaknanda river, the land under river valleys is suitable for cultivation. About 75 per cent arable land is recorded between 600 m and 1800 m height m.s.l. The land above 1800 m height from m.s.l. is richly favourable for horticulture, millets and potatoes. In the district of Chamoli only 55800 hectares of land is under agriculture, i.e., 6.11 per cent because more than 50 per cent of the northern part of the district is mountainous, clothed with forests and alpine pastures and snow. In the districts of Pauri and Tehri, the basin occupies 30878 and 28841 hectares of land under agriculture respectively out of the total area.

Table II
Land Use Pattern

Major landuse categories	Landuse sub-categories		Area (km^2)	Percentage
1	2		3	4
			1148.19	10.59
1. Cultivated	(i)	Cropped land	700.445	6.43
	(ii)	Fallow land	447.745	4.16
2. Forest			3078.62	29.87
	(iii)	Dense forest	1989.62	18.28
	(iv)	Sparse forest	1089.00	10.03
3. Cultivable waste	(v)	Cultivable waste	170.55	1.56
4. Barren land	(vi)	Barren land	2967.66	27.25
5. Pasture land	(vii)	Alpine pasture	773.52	7.10
6. Land under glacier and snow	(viii)	Land under glacier and snow	2504.99	23.13
7. Miscellaneous	(ix)	Miscellaneous land	238.00	
	Total Area km^2		10880.16	100.00

Forest Land

The large part of the basin is covered by forest land. Of the total area of the basin about 324800 hectares of land are available under forest i.e., 29.87 per cent. The total lend under forest is further sub-divided in dense and sparse forests. The area under dense forest is 1989.62 hectares i.e., 18.28 per cent and rest of 145900 hectares are included in sparse forest forming 11.59 per cent of the total area of the valley.

Cultivable Waste

Land under cultivable waste consists of 17055 hectares of

the total area of the basin i.e., 1.56 per cent. Of the total cultivable waste land, 3402.90 hectares occur in Tehri district and 3202.10 hectares in Pauri district and the rest of 10496 hec tares in the district of Chamoli. The said land comprises of scrubs and bushes and thick cover of soil patches under gullies.

Barren Land

Next to forest land, about 2967.46 hectares are recorded as barren land which forms 18.59 per cent of the total area of the basin. Of the total area of the barren land 2300 hectares are found in the district of Tehri and 7808 hectares in Pauri district and the rest of 286638 hectares are recorded in Chamoli district. The barren land generally includes rocky, stoney and deeply dissected mountainous land with summits. The area of barren land is increasing gradually on account of deforestation, over quarrying, over grazing and other excess of human interference, which need ecological measures for immediate treatment.

Land under Glacier and Snow

Of the total area of the Alaknanda Basin, 250499 hectares of land occur under glaciers and snow forming 13.614 per cent of the total land of the basin. Almost the land under glaciers and snow forms the northern part of Chamoli districts and a few patches of 5000 hectares are in Pithoragarh and Almora district of Kumaon Mandal forming the eastern limit of the basin.

Land under permanent pastures

Permanent pastures occupy 7.10 per cent area of the basin i.e., 773.52 hectares. These pastures are found between 3300 m and 2900 m height in north and north-eastern part of the Chamoli district. They are called Alpine pastures. In the local dialect they are known as Payars and Bugyals. The grasses of these payars and Bugyals are very nutritious for sheep and goats.

Miscellaneous land

The miscellaneous land which includes the area under settlements, roads, rivers, lakes and ponds, etc. consists of 238.00 hectares i.e., 2.06 per cent of the total land of the basin.

CROPPING PATTERNS

Cropping pattern of the valley primarily follows the nature

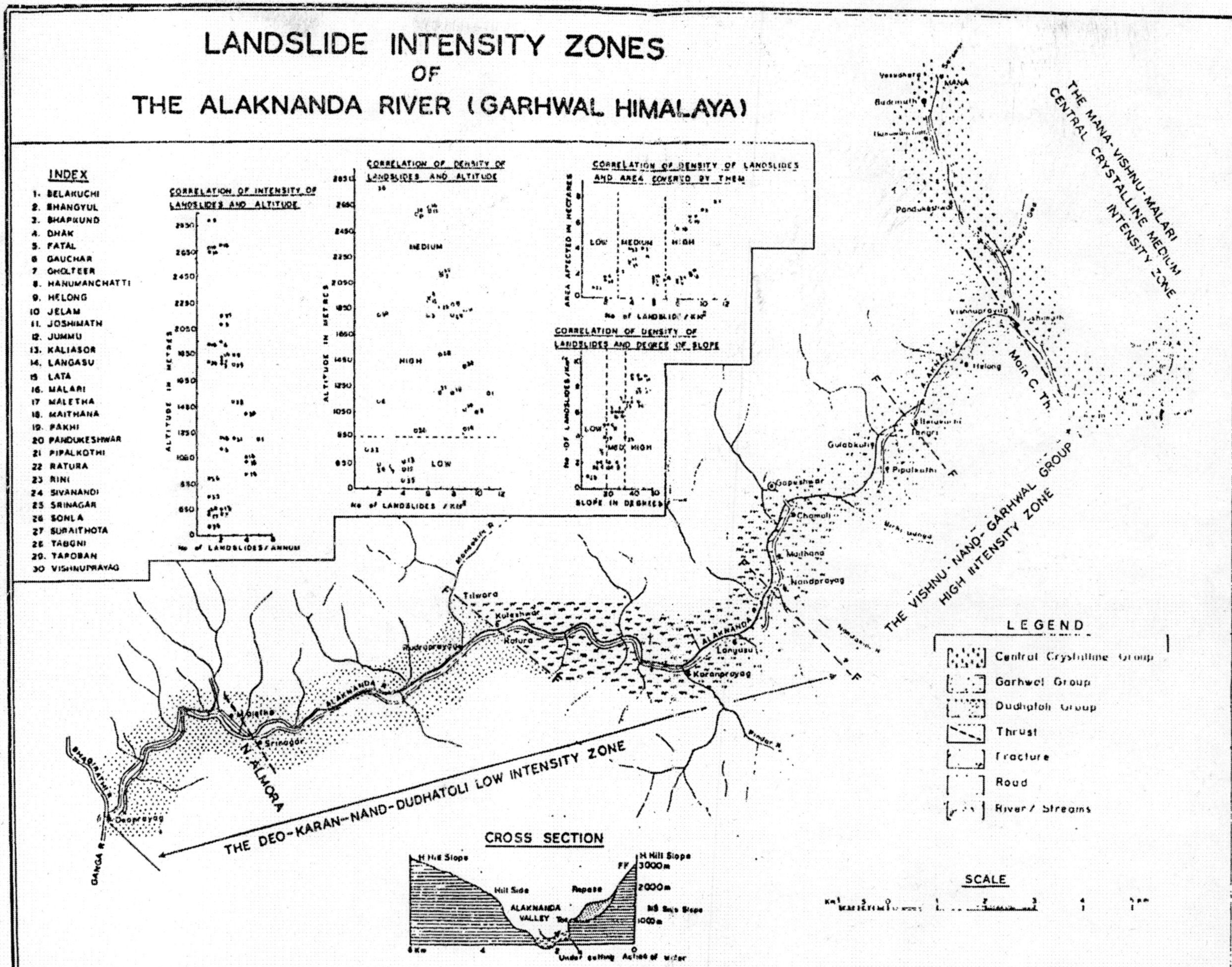

Fig. 4: Land-slide-Intensity Zones of Alaknanda River (Garhwal Himalaya)

(Facing page 26)

of physiography, climate, soil and aspect of landforms. Agricultural practices and cropping system persist to wrest with severe water and soil depletion hazards, increasing population d decreasing productivity of soils. Almost both the rabi and rif crops are harvested in the valley.

pping Pattern in Rabi Season

The rabi season generally continues from September and Oc-r to March, April, May and even in July. The span of the son varies according to the change in altitute. In the higher titude wheat, which is prominent and widespread crop is harested in the month of May at an altitude of 1800 m, in June at n altitude of 2100 m and in July at 2400 m from m.s.l. Main crops in this season are wheat, barley, pulses, oil seeds, vegetables and potatoes. It is generally recorded that major portion i.e., 60 per cent of the cultivated land is used under rabi crops. Wheat and barley are important crops of the basin. Of the total area under rabi crops i.e., 56215.37 hectares of the Alaknanda basin about 214600.00 hectares of land was occupied in Chamoli district, 24252.97 hectares in Pauri district and 10502.40 hectares in Tehri district in the year 1966.[1]

Wheat

Wheat is the main crop of the rabi season in the Alaknanda valley. It is harvested upto 3000 m height from m.s.l. The total area under wheat crop in the valley was recorded 31149 hectares which occupied 60 per cent land of the area under rabi crops. Generally wheat is sown in every block of the Alaknanda basin. The maximum area under wheat crop occurs in Devprayag (7208 Hectares), Augustmuni (3403 Hectares) and Kot block (2398 Hectars) in the valley but the highest yield i.e., 5000 Kg/Hectares was recorded in Augustmuni block due to having better means of irrigaiton and thick cover of soils. The minimum land under wheat crop recorded in Joshimath i.e., 95.60 Hectares and Ukhimath i.e., 1400 Hectares. The minimum yield of wheat was obtained in Joshimath and Ukhimath block i.e., 492 Kg/Hectares respectively due to lack of irrigation, thin cover of soil and poor management. The rate of productivity is still the same.

Barley

Barley is the second main crop of the rabi season in the valley because it may be harvested in low rainfall area and low temperature of the high altitude ranging upto 3308 m and it takes only six months period for harvesting. In the Niti and

Mana villages of the Upper Alaknanda Basin, barley, owa, kauni and fafar are most important crops. 'Tao' is another variety of barley which is common in the upper and Middle reaches of the Alaknanda Basin. The total area under barley was recorded as 7635.13 hectares in the year 1966, but now-a-days the area under barley is being reduced gradually because farmers take much interest to grow wheat on account of some initiatives given to them by the Government. The maximum area under barley is recorded in various blocks like Devprayag (1041.60 hectares), Kot (1072.02 hectares), Gairsain (690.40 hectares), Narain bagar (655.60 hectares) and minimum area under barley is recorded in Tharali (151.20 hectares) and Nagpur (189.60 hectares) blocks where barley is being gradually replaced by wheat due to availability of better means of irrigation and Government policy in favour of wheat. The maximum yield per hectares is 1500 Kg. as recorded in Augustmuni block and lowest yield per Hectare occurs in Nagpur block i.e., 189.60 Kg.

Oil seeds and pulses

Oil seeds and pulses are the secondary crops in rabi season. Almost in every block pulses and oil seeds are harvested but they have low rate of yield i.e., 900 kg/hectare in the valleys. The total land under oil seeds is recorded as 189.60 hectares in the rabi crops. The maximum area lying under oil seeds occurs in Augustmuni i.e., 90.00 hectares and minimum land under oil seeds occurs in Gairsain i.e., 16.80 hectares due to lack of favourable conditions for oil seeds.

The total area under pulses in the valley was recorded 3200.80 hectares in 1966. Generally maximum land under pulses occurs in Jakholi i.e., 1485 hectares in Devprayag (686 hectares). The minimum land under pulses occurs in Tharali i.e., 1.66 hectares and in Gairsain i.e., 1.65 hectares. The maximum land under pulses is recorded in Narain bagar and Nagpur i.e., 40.00 and 41.60 hectares respectively.

Vegetables and potatoes

Vegetables are sown in a few blocks in the valley. The total area under vegetables in the valley was recorded as 285.71 hectares. Out of which 189.71 hectares were available in Pauri, 42.80 hectares in Ukhimath, 17.20 hectares in Gairsain, 40 hectares in Jakholi and 30 hectares in Kirtinagar and rest of 20 hectares in Devprayag. Thus the maximum land under vegetable is available in Pauri, Ukhimath and Gairsain blocks. Area under vegetables may be extended on the lowest terrace of the river valley where irrigation facilities are available. The cultivation of vegetables may be more economic in the vicinity of pilgrimage towns like Vishnuprayag, Nandprayag, Karanprayag, Rudraprayag and Devprayag. In general, the vegetables are supplied from Kotdwar and Hardwar hence fresh vegetables may not be available everywhere.

Potatoes

In the valley, potato is harvested in both rabi and kharif seasons. Total area under potatoes in rabi season was recorded 242 hectares in 1966. Potatoe is harvested in Ukhimath block (145 hectares) and Augustmuni block (13.20 hectares) in the Chamoli district, Jakholi (87 hectares), Devprayag (101 hectares) and Kirtinagar (20 hectares) in the part of Tehri district of the valley. The total area under potato in rabi season may also be extended in Ukhimath block. The yield of potato in Ukhimath Tehsil is 600 quintals per hectare. This is the most economic crop in the valley. The area and yield of potato may be increased if irrigation and marketing facilities are provided by the Government.

Other crops

There are a few more coarse grains grown in rabi season between 1800 m and 3300 m height. The coarse grains like kauni, owa, fafar, jai, ogal are generally harvested in the Dhauli and Kedar valleys in the upper part of the Alaknanda basin. The total area under the said coarse grains is recorded as 9487 hectares. The cultivated area of the said crops is very much infertile, uneven and unirrigated. The yield may be increased if means of irrigation are extended in remote areas of the river valleys.

Cropping Pattern in Kharif Season

In the valley kharif season continues mostly from April, May and June and harvesting period occurs in the months of September, October, November. The main crops in kharif season are paddy, madua, jhangora, ramdana and potatoes. In the valley, where irrigation facilities are available paddy is grown.

The total land under kharif crops was recorded as 5858.51 hectares in 1966. Of the total land under kharif crops 60 per cent was under paddy and rest of 40 per cent was under madua, jhangora, ramdana, vegetables and potatoes.

Paddy

Paddy is a primary crop in the kharif season. The cultivation of paddy is being carried between 450 m and 1800 m height from m.s.l. The cultivation of paddy is performed under following systems:

(i) Ukhar (rainfed) paddy cultivation.
(ii) Sera (wet) paddy cultivation.

In the system of ukhar paddy cultivation, the rice is generally sown in the month of January-February in the valleys of higher altitude upto 1800 m height from m.s.l. where irrigation facilities are not available. Such type of land under paddy in the Alaknanda valley is called Ukhar land. Of the total area under paddy, 40 per cent land is recorded under Ukhar paddy and rest of 60 per cent under Sera paddy.

In the Sera system of paddy cultivation, the paddy is sown in the months of June and July and it is harvested in the months of September and October. The average yield of this paddy in valley is 20 quintals per hectares. The highest yield i.e., 30 quintals per hectares is recorded in Augustmuni block. The total land under both the types of paddy was 21929 hectares. Of the total area of rice, about 2520 hectares are available in Pauri block, 2803.2 hectares in Khirsu and 2402 hectares in Kot, 3370.80 hectares in Augustmuni and 2377 hectares in Narainbagar possessing high yield ranging from 20 to 30 quintals per hectare. The rest of the blocks in Alaknanda basin have low yield i.e., less than 20 quintals per hectare due to lack of means of irrigation and thin layer of soil like Joshimath. Thus Augustmuni, Narainbazar, Pauri, Kot blocks are famous for rice production in the valley. Besides these, Phata Nagpur, Simli and Ghultur localities have also been developing gradually.

During five year plan, the Government introduced the Japanese method of rice plantation but under experimental stage.

Madua

Madua is one of the coarse grains in the Alaknanda valley. It is mostly produced in the poor quality land, where means of irrigation are not available and soil has low fertility. Madua is harvested upto 2100 m height from m.s.l. in the valley. The total land under madua was recorded 17815.16 hectares in the year 1966. The maximum land under madua is recorded in Augustmuni block (2317 hectares) in the district of Chamoli, in Pauri block (1700 hectares) in the district of Pauri and Kirtinagar block (2606.40 hectares) in the district of Tehri covering Alaknanda valley partly. The minimum land under madua i.e., 34.80 hectares is recorded in Ukhimath block in the district of Chamoli and in

Khirsu block (30.76 hectares) in Tehri district comprising the Alaknanda basin partly.

Jhangora

Jhangora is another popular coarse grain which is generally sown in the medium quality of land. It is cultivated in every block like madua crop of the total land of kharif season. Jhangora occupies 8188 hectares of land i.e., 69 per cent in the valley. Maximum land under Jhangora is available in Ukhimath block (1162.80 hectares) in the Chamoli district and in Devprayag block (1938.40 hectares) in the Tehri district covering the basin partly. The minimum land under jhangora i.e., 88 hectares land is recorded in Dasoli block in Chamoli district and 94.8 hectares in Jakholi block in Tehri district.

Potatoes

The total land under potato crop in the season was 1460.40 hectares i.e., 2.6 per cent of the total land of kharif crops. The maximum land under potatoes in the kharif season i.e., 400 hectares is recorded in Joshimath block and minimum land under potatoes i.e., 24 hectares is recorded in Gairsain block due to lack of means of irrigation. Generally potato is produced in the whole basin. Its cultivation may also be extended if means of irrigation are developed. Potatoes may be produced upto 3300 m height from m.s.l.

Other crops

In the valley a few crops like maize, pulses and green vegetable are also produced. The total land under maize was 1446.65 hectares (25 per cent). Maize is mostly harvested in Jakholi, Kirtinagar, Devprayag, Pauri, Khirsu and Kot blocks in the valley. The maximum land under maize i.e., 225 hectares is recorded in Pauri block and minimum land i.e., 14 hectares is available in Kirtinagar block.

As regards the cultivation of pulses in kharif season, the total land under pulses is 1994.4 hectares in the year 1966. In pulses tur is mostly harvested in all blocks of the basin. The maximum land under pulses i.e., 638.80 hectares is recorded in Dasoli and minimum land i.e. 8.4 hectares is recorded in Kirtinagar. The area under pulses like tur may also be extended in the cultivable waste land.

Besides these the total area under vegetable in kharif season was 299.20 hectares in 1966. Vegetables are generally harvested in Jakholi, Devprayag, Pauri, Khirsu and Kot blocks in the valley. The maximum land under vegetables was recorded 193.20 hectares in Devprayag. The land under vegetables may also be increased if means of irrigation and marketing facilities may be provided by the Government.

AGRICULTURAL SYSTEM

The nature has endowed the valley with glacial, glaciofluvial and inter-montane fluvial landforms, green vegetation, snow-fed rivers, sulphur-springs, soils and minerals. Even its rural life primarily based on agriculture, is underdeveloped economically on account of the multiplying results of the resource deplection factors like soil-erosion, land-slides, soil creeping and avalaunching, which occur due to over-grazing, unscientific quarrying, torrential run-off, severe infiltration, bare-fallowing and over-blasting for construction works and deforestation. The above depletion of land resources in the basin occurs due to the adverse effect of the climate and the neglect of the resource conservation measures by the people. Under such adverse conditions, the best suited modern agricultural set-up has been evolved after a long experience. Generally, the farming techniques, crop-rotation, crop-association, field patterns and the duration of crop sowing, growing and harvesting in the valley differ with the change of altitude in the basin and the altitude governs the climatic conditions therein. The climate plays a significant role to delimit the zones of different crops on different altitudes, slope and aspects. It is usually observed that rice and jhangora (Panicum frumentaceum) may grow below 1850 m, while marsa (Amaranthus frumentaceum) grows below 2750 m and fafar (Faopyrum tataricum) may grow upto 4200 m (Table IV).

In the hilly terrain like Garhwal Himalaya, temperature is more important phenomena to determine the limit of crops, than rainfall (just reverse to the plains where rainfall delimits the crop zones because temperature is available in sufficient amount for ripening of crops). In the Garhwal Himalaya, where rainfall is recorded 250 mm to 2000 mm sufficient for Himalayan crops, but temperature decreases with the increase of altitude, which permits rice to grow upto 1800 m and marsa (Amaranthus frumentacoum) to 2700 m. The crop harvesting period also differs with the change of temperature. For example, wheat is harvested in the month of May at the altitude of 1800 m in June at 2100 m, in July at 2400 m and in August at 2700 m and above. The temperature also controls, duration and amount of snowfall, which influence the period of sowing, growing and harvesting of crops on varied altitude of pereglacial areas.

Similarly soil is very essential component of the ecosystem for studying the agricultural system of any region. In the Lower Alaknanda Basin, less clayey loam, acidic and gravelly soils contain pH values from 6.00 to 7.00. However, in many villages, located on the lower benches of the river valley, alluvials of clayey loam have been noticed in broad patches in Maletha, Nagrasu, Gauchar and Langasu (Table III).

In the Middle Reach of Alaknanda Basin, the podzolic soils are leached and poor in fertility. The pH values vary from 6.0 to 6.6 suitable for gardening of apples and potato cultivation. In the Upper reach of the Alaknanda Basin, the soils are stoney, glacial and acidic. pH values go below 6. Such infertile soils is suitable for cold-resisting crops like fafar, barley, chua, cheena, ogal, potato and hard wheat etc. (Table V).

Besides above components of the ecosystem, the aspects of slope (northern aspect and southern aspect) also plays an important role primarily in the hilly terrain to favour and disfavour the growth of tree crops. The southern aspect (sunny slope) is more favourable for ripening of crops than that of the northern aspect (shady part of the hill slope) which is highly suitable for long period moisture-retaining tree crops (forests).

Thus the ruggedness of the hilly terrain of the Alaknanda basin and its micro-diversity in climatic conditions, have evolved varied systems of tilling, crop rotation, crop association, crop intensity, field morphology and in irrigation devices as detailed below:

Size slope and morphology of agricultural fields

Generally the shape, size and morphology of the fields are coined primarily by the nature of physiography and drainage. In the Upper and Middle Alaknanda Basin, mostly the inter-locking spurs have been used for field terraces with great toil. The river cut-benches are very narrow, elongated and irregular. The average size of field terrace in the Upper Basin may range from 5 to 15 metre long and 1 m to 5 m wide and 10 m to 25 m long and 1.5 m to 7 m wide in the middle reach of the valley.

The annualar, triangular, rectilinear and irregular shaped terrace fields are found in the upper and middle reaches of the valley. Generally annular and triangular fields are observed where river bends and confluences exist, for example, the fields of Maletha, Nagrasu and Paperasu, Dhari, Kirmoli and Kothi. Rectilinear fields are found where valley becomes very narrow, deep and straight as in the villages of Rilni, Dhari and

Kaliasaur. Annular formations of fields also repeat on the break of slopes of the spurs as in the villages of Paprasu, Dungri-Pant and Frasu. Annular formations of field are of two types: (1) annular with convexity, and (ii) annular with concavity. Both are observed on convex and concave sides of the interlocking spurs in the upper reach of the Alaknanda and its tributaries. Fan-shaped fields are generally formed on the glacial drifts and major landslide drifts as observed in the Malari, Lata and Kailaspur in the Dhauli valley.

Broad parallel alluvial terraces are formed in the Augustmuni, Chandrapur (Kedar valley), Gauchar, Nagrasu and Langasu, where the valley has been broadened and paired terraces have been formed. These terraces fields are very wide, thick and long. Rock benche fields are also observed in the valley near Belakuchi, Barhgaon, Jelum and Pandukeshwar where one side of the valley is steep and otherwise consists of slip off slope. Field terraces generally lie both above and below the villages which are usually situated on the mid slope of a spur. The land of the upper terrace is called "upraon land" and of lower terrace 'talaon land'. The upraon land is usually found unirrigated (rainfed) in which koda (Paspalum miliaceum), madua (Eleusine coracana), Chenna (Panicum miliaceum) and fafar (Fagopyrum tatarcum) are harvested.

In the talaon land fields may be irrigated by gul system to harvest potato, wheat and hill rice. The unirrigated and unterraced old fallow is called "Katil" land. On the mid slope of the high spurs, the terrace fields are so narrow that plough may not be used and fields are tilled with a hoe called "kutla" in the local dialect. In the broad terraces of the lower reach of the Alaknanda, fields are well ploughed and also irrigated with canals and kuls.

Cropping pattern

In all the ecological sub-regions of the valley, certain sets of crops (in rabi and kharif season) are harvested. Rice, jhangora, madua, maize and wheat, etc. and legume crops are also grown in the Lower Alaknanda Basin below 1200 m. In the middle Alaknanda Basin, rice, madua, cheena, kauni, wheat, barley, potatoes, kulath, chaulai, etc. Wheat, barley, potato, fafar, uva, ogal, marsa are harvested in the Upper Alaknanda Basin. Out of the above crops, wheat, potato, madua and barley crops are mostly grown in every zone. The cultivation of apples, pears, citrus fruits, cherries, plums, walnut and apricots are suitable in the zone ranging from 1200 to 2400 m. In the basin, mostly crops are grown separately, but a few like madua, dry paddy and wheat are sown with their related crops. It is generally observed that madua + kulath + bhat + urad are grown together in the same fields. In the fields of rainfed paddy, the madua is sown on the buds of the fields. Wheat is also sown with mustard and peas. Third set of jhangora is associated with kauni. Barley is grown separately. Phapar, uva, cheena, marsa are also harvested separately.

Crop Rotation

Not only in the Alaknanda Basin but in the Himalayan Region, the existing crop-rotation cycles have been evolved after a long experience. They are well suited to such a terrain but now irregular and haphazard land use and the neglect of land resource conservation measures reduce the crop yield in the valley under certain crop rotation. As per personal investigations the following crop rotations prevail in the Alaknanda Basin:

S.No.	Cycle of crop rotation	Code
1.	Cycle of 1 year two crop rotation	($Cy^1$2CR) Lower/Middle Alak
2.	Cycle of 2 year three crop rotation	($Cy^2$3CR) Lower/Middle Alak Sar System
3.	Cycle of 2 year three crop rotation	($Cy^2$3CR - Middle Alak)
4.	Cycle of 5 ye͵r seven crop rotation	($Cy^5$7CR - Upper Alak)
5.	Cycle of 5 year three crop rotation	($Cy^5$3CR - Remote Lands Upper Alak)
6.	Cycle of nine year three crop rotation	($Cy^9$3CR - Kotil) Upper Alak.

(1) <u>Cycle of one year two crop rotation</u> ($Cy^1$2CR-Lower/Middle Alak)

Generally this crop rotation prevails in the talaon land (low land) which is well irrigated in the lower and middle Alaknanda basin. Irrigated terraces on the talaon land are generally extended towards the valley bottom. In such fields, two crops are harvested each year without leaving land as fallow. Wherever, the fields are not adequately irrigated, jhangora and wheat crops are harvested. This type of rotation prevails mostly in the lower Alaknanda basin.

(2) Cycle of 2 year three crop rotation ($Cy^2$3CR - Sar System - Lower/Middle Alak)

This type of crop rotation usually is practised in the upraon land where fields are unirrigated (rainfed) and no means of irrigation are available. The cycle of said crop rotation follows a particular sar system. Generally in this system farmers divide their field in two components (sars). In the first year, one half of the field grows rice and the other half grows madua (from April to September) pronounced as "satiyara" and "kodara" respectively in the local dialect.

Both the satiyara sar and kodara sar have alternate crops, using fallow devices timely to utilise the maximum gain of carrying capacity of land as noted below:

Sar System

1st Sar (Rice Sar)		2nd Sar (Kodara Sar) (Madua Sar)
Crop	Period	Crop
Rice	April-September	Madua
Wheat	Oct.-April	Fallow
Madua	April-September	Rice
Fallow	Oct.-April	Wheat
Rice	April-September	Madua

The above crop rotation indicate that wheat and barley may not be sown just after madua because the land exhausts its carrying capacity and the land becomes less fertile for wheat production. This cycle of sar crop rotation prominently prevails in the Garhwal Himalaya on upraon land in the lower and middle reaches of the Alaknanda river.

(3) Cycle of 2 year three crop rotation ($Cy^2$3CR - middle Alak)

In the Middle Alaknanda Basin, the usual practice of two year of three crop rotation prevails. The duration of wheat in this zone is longer than that of lower Alaknanda basin (Table V). The said cycle of crop rotation goes as:

	Crops	Period		
1.	Marsa or ogal or fafar	April - Sept.	One year	
2.	Wheat or barley	Oct. - June		
3.	Potato or mustard or beans	Aug. - dec.	One year	Two year
4.	Fallow	Dec. - April		

This is a usual practice in this zone that in the year in which marsa or ogal is sown on the field, potatoes or beans are grown on the other field and in the next year, field exchange their crops.

(4) Cycle of 5 year seven crop rotation $Cy^5$7CR - Upper Alak)

In the Upper Alaknanda Basin, the soil is mostly inefficient and means of irrigation are also not properly available. Mostly rainfed crops are harvested with the result farmers have to leave the land for three times fallow in the completion of each cycle of five years as noted below :

S.No.	Crops	Duration	
1.	Potatoes	1st Summer	1st year
2.	Wheat or barley	1st Winter	
3.	Madua	2nd Summer	2nd year
4.	Fallow	2nd Winter	
5.	Potatoes	3rd Summer	3rd year
6.	Wheat or barley	3rd Winter	
7.	Fallow	4th Summer	4th year
8.	Wheat or barley	4th Winter	
9.	Marsa or kauni or ogal	5th Summer	5th year
10.	Fallow	5th Winter	

(5) Cycle of 5 year three crop rotation ($Cy^5$3CR - Remote Lands - Upper Alak)

On the outskirt and remote land of the villages in the Upper Alaknanda Basin, above cycle of crop rotation prevails, because such infertile land seems to be very poor, marginal and unirrigated. Such type of land has been reclaimed and encroached from the forest fringe land. Generally the farmers follow the fallow device for one season after harvesting two crops from the field continuously then again the same field is left as fallow for coming three years after harvesting third crop. The system is explained below:

S.No.	Crops	Duration	
1.	Wheat or barley	1st Winter	1st year
2.	Kodo, kauni or marsa	1st Summer	
3.	Fallow	2nd Winter	2nd year
4.	Jhangora or marsa	2nd Summer	
5.	Fallow	Three years	3rd to 5th year

6. Cycle of 9 years three crop rotation ($Cy^9$3CR - Katil land - Upper Alak)

The "Katil" land is very dry, unirrigated, unterraced reclaimed land from the forest fringe. It is reclaimed by burning a patch of forest. The land is found so poor that it regains its fertility after leaving it as fallow for many years. The system of crop rotation goes as:

S.No.	Crops	Duration	
1.	Kodo or fafar	1st Summer	1st year
2.	Barley or wheat	1st Winter	
3.	Fallow	2nd Summer	2nd year
4.	Fallow	2nd Winter	
5.	Ogal or fafar	3rd Summer	3rd half year
6.	Fallow	3rd Winter onwards	3rd to 9th year

In the Dhauli valley, the potato crop is followed by fafar as investigated in the villages like Jalam and Malari. The farmers follow crop rotation like:

Potatoes	..	Fafor	..	March/April - Oct/Nov.
Fallow	..	Fallow	..	Due to cold and snow
Faforq	..	Potatoes	..	March/April - Oct/Nov.

Irrigation

In the villages of Jelam, Malari, Kailash, Bampa, which are settled on mid slope of the spur or glacial drifts, at the altitude, ranging from 8000' to 12000', the Bhutias harvest two crops yearly due to the provisions of a few kuls in the valley. There are about six kuls (canals) in the Dhauli valley which are not still sufficient to irrigate the cultivated land. These are:

1. Malari kul
2. Bampa kul
3. Kailash kul
4. Furk kul
5. Gansali kul
6. Niti kul

Horticulture

According to climatic and topographical conditions the horticulture is ideally suited. The region is altitudinally varied from 450 m to 4500 m height from m.s.l. Nature and kind of fruit tree crops vary according to the change of altitude. There are three types of fruits orchards as discussed below:

(1) Mixed fruit belt

The belt comprises of mango and guava fruit gardens, which are found in between 450 m and 1200 m height because these fruits require tropical and sub-tropical conditions which prevail in lower river valleys of the region.

(2) Citrus fruit belt

The belt ranges from 12[illegible] to 2100 m height where temperte and sub-temperate conditions prevail freely. Such climatic conditions are ideally suited for citrus farming as in California (U.S.A). The dominant citrus fruits are orange, lemons and chakotra besides these fruits plum, khumani and cerry are also produced there.

(3) Chestnut, valnut, local almost dominant fruit belt

The belt prevails between 2100 m and 2200 m height from m.s.l. The chestnut and valnuts are produced in this belt. This belt may also be extended in the higher altitude.

Horticultural Development in Five Year Plan

Before 1950 people were not well known with the horticulture in the valley. At present there are various nurseries and orchards as given below:

Name of nursories	Area under nurseries (Hectares)
Parsari	21.20
Subhani	3.20
Singhdhar	2.00
Gwaldam	4.60
Jakhdhar	12.80

Under the scheme of five year plans, horticulture mobile teams were established in every block of the valley to help the work and to promote horticulture on sound footing. The teams were made to get the farmers awarded of the knowledge of the modern techniques of horticulture.

Achievements during five year

Achievement	Area under III Five Year Plan	Area under IV Five Year Plan
1. New orchards, plantations in hactares	312.00	1647.20
2. Distribution seedlinge of fruit trees	411805	582500
3. Production of vegetable seeds(kg)	557.09	1000.00
4. No. of small orchards	1487	2000

Besides these, the major scheme has also been considered to develop along the motor roads. In the part of the valley lying in Tehri district there have been marked three fruit belts as given below:

1. Chirbatia Panwali fruit belt (Jakholi block - 24 km
2. Hindolakhal to Chandrabani fruit belt - 16 km
3. Sonar to Khencha (Kirtinagar block) - 27 km

In the upper and middle Alaknanda valley, mostly lying in the Chamoli district there records about 706 private and 55 Government Orchards which occupy 1133.80 hectares of land comprising of 1189545 trees as given in Table II. The government have given various incentives to the farmers of the region to develop orchards but it has been noticed that the loan and subsidies provided to the farmers for fruit plantation and for the purchase of livestocks have been misused by more than 60 per cent. The author has personally interviewed the farmers with their families at Sabdharkhal, Dewalgarh, Thalisain, Rikhanikhal and found that the farmers have made loan system as the source of income and they have lost the consciousness towards the development of the region. Even educated people like teachers of the government Inter Colleges of the said localities also confirmed the causes of the failure of the said scheme. In a real sense the farmers are pressurised to do so by the authorities who involve in sanctioning the loan and giving subsidies.

Hence the Government should change the policy and each

and every tree of the orchards should be numbered and a few norms should be laid down to survive trees upto a certain height through a committee of village farmers who will be responsible for all consequences. The said scheme has also been considered for the development of apple plantation and vegetables along the motor roads.

Table III
Distribution of orchards in Alaknanda Basin

Blocks	Total Land under orchards (Hec.)	Private orchards (No.)	Govt. orchards (No.)	Total No. of fruit trees
Joshimath	160.00	350	27	889161
Narain-bagar	170.00	-	-	647080
Gairsain	142.80			32500
Karan-prayag	46.00			3500
Ukhimath	70.00			41720
Augustmuni	169.20			90824
Tharali	104.00	246	10	96132
Nagpur (Pokhari)	154.00	110	18	50663
Chamoli	118.00			47025
Total	1133.80	706	55	1189545

Source: District Horticulture Office, Gopeshwar, Chamoli.

Live Stock and Pastoralism

The people of Chamoli district are mainly pastoralists and farmers. Animal husbandary and pastoralism are the important aspects of the development of economical condition of the people. Due to break of India-Tibetan trade, now it has been very essential to pay maximum attention to this field for its regional development. The domestic animals of the region are generally weak. The farmers do not care and do not feed them potential foddar.

Generally people do not take much interest in agriculture due to lack of means of irrigation and other agricultural facilities hence the conditions of the livestock is also deplorable in the valley.

Prospects of Animal Husbandary

Development of sheep and goats

Half of the people of the region are engaged in this occupation. Most of the part of the region, which are more than 1800 m high from m.s.l. offers favourable climate for sheep and goats. About 125017 sheep and 95882 goats were enumerated in the year 1071.

Poultry development

In the year 1966 there were about 6703 cocks and hens in the region. Now block development officers are providing loan facilities to develop poultry in the region. But it is not developing well because of disfavourable climate conditions of the basin.

Fisheries

Below 1200 m altitude the fisheries get favourable conditions. Alaknanda and its tributaries like Mandakini, Nandakini, Pindar rivers are very suitable for fishing and fisheries.

Delimitation of Agricultural Crops

The region may be very much dissected and uneven, finds various limitations for agricultural activities. There have been worked out three altitudinal cereal limits on the basis of ecological parameters given in Table V.

1. Low mountain wheat-paddy dominant cereal limit

   (450 - 1200 m) $\frac{HHM}{111} - \frac{Hyad}{112}$

2. High mountains madua-Jhangora-dominant cereal limit

   (1200 - 2400 m) $\frac{Hgr}{211} - \frac{Hyad}{212}$

3. Very high mountain fafar, ogal, kauni, potatoes dominant cereal limit (2400 - 3300 m).

   $\frac{Lwt}{321} - \frac{Ahd}{312}, \frac{Lwt}{324} - \frac{TEW}{424}$

(1) Low mountain wheat, paddy dominant cereal limit

About 50 per cent cultivated land is lying in this limit in which summers are hot and winters are cold. Mean annual temperature ranges from 12^{0} to 21^{0}C and annual rainfall occurs 2000 mm. Pedologically the zone lying in between 450 m and 1200 m comprises of thick layer of good soils with its certain properties like pH value ranging from 7 to 8 E.C./mm hos from 10 to 14, organic matter from 8-12 per cent and nitrogen 40 per cent and the crop gradient pattern based on photosynthesis as (Purohit, 1971) + s, c^4 indicating subtropical climate. Thus the zone in which the said parameters are absolutely favourable for wheat and paddy, delimits the area of the Alaknanda basin upto Chamoli in the north and Gwaldam in the east and Fata in the north-west. The zone is also enriched with canal irrigation systems. Mostly the crops are harvested in the Talaon and upraon lands of the river valleys of the low mountains. $\frac{\text{Hyr}}{\text{III}}$(fig.)

(2) <u>High mountains madua-jhangora-dominant cereal limit</u>

The zone consists of 40 per cent cultivated land between 1200 m and 2100 m height from m.s.l. Climatically the zone experiences subtemperate conditions. The average annual rainfall varies from 1050 mm to 2000 mm. The mean annual temperature ranges from 10^{0}C to 13^{0}C. Pedologically the zone consists of sandy and sand loam with gravel (33 per cent) pH value ranges from 6 to 8 per cent. The organic matter in the soil varies from 4 to 8 per cent, nitrogen varies from 20 to 40 per cent. Crop gradient based on photosynthesis is + S + CAM. S + Cam + c^4 is ecologically suited for growing millets, Jangora, maize and potatoes. Wheat and paddy are also harvested but their yields is very low which may also be increased if means of irrigation and fertilizers are provided timely. $\frac{\text{Hyr,}}{221}$ $\frac{\text{hyad}}{212}$ (fig).

(3) <u>Very high mountain Fafar, Ogal Kauni dominant cereal limit -</u>

The zone consists of 10 per cent of cultivated land between 2400 m and 3300 m height from m.s.l. Of the total land of the zone 75 per cent land is completely barren and sparsely forested. Only 25 per cent land is available for agriculture. $\frac{\text{LWH}}{312}$(fig)

Climatically the zone experiences temperate to cold conditions, with mean annjual temperature varying from 8^{0}C 10^{0}C and mean annual rainfall ranging from 250 mm to 1050 mm. Snowfall occurs during winters above 300 m height from the sea level. $\frac{\text{TLW}}{424}$(fig.)

Pedologically the terrain of katil land consists of sandy soils with gravels and pabbles (33 per cent). Thus the zone due to being dissected, uneven by fast flow of runoff of local torrents, seems very much infertile. The gravelly soils of the zone comprises of low organic matter i.e. 4 per cent E.C. mmhos less than 10 and pH meter less than 6 and nitrogen less than 20 per cent. The crop gradient pattern based on photosynthesis i.e. (C^3 + 5 + cm + 5 + cam + C^4) depicts ecological conditions for fafar, orgal, kauni and potatoes cereals.

After analysing the significant relationship among physiography (relief, aspect and slope) climate, soils, gradient of photosynthesis pathways and landuse, the following regional land forms have been demarcated to diagnose the intensity of relational ailments of land resources and also to evaluate their land capability of different regions for recommending ecological treatments in the valley for its rational agricultural landuse planning. Broadly the landuse pattern land under low mountain range is depicted by Hym/111 fig. and forest land by D9/112. Similarly the cultivated patches forest open scrub and barren lands have also been shown under varid fractional codes as Hym/221 (cult) Hyad/212 (forest) Lwt/311 (cult) Ahd/312 (forest) Lwt/324 (Barren) and Lwt/424 Fig.

In the Alaknanda Basin the broad land use relationship with soils, physiography subunits and aspect has been demarcated at least in three digits in numerator and three digits in denumerator of the resultant fractional code. The said relationship with cultivated land, forest, open scrubs and barren land in different physiographic subunits has been indicated under varied fractional codes as Hym/111 (cult). D9/112 (forest), Hym/211 cult, Hyad/212 (forest), Lwt/311 (cult) Ahd/312 (forest Lwt (barren) (Fig.4).

The study of detailed relationship with physiography subunits, aspects, slope, climate and groups subgroups and family of soils, have been conducted in the Lower Alaknanda Basin (The Augustmuni block).

Reference

1. The land use data for the year 1966 were collected to compare the landuse figures worked out (from land sat. imageries and photographs of the same year) by remote sensing techniques to obtain accuracy in landuse mapping.

Table IV
Altitudinal Distribution of Crops Grown in the Alaknanda Basin

S. No.	Local Name	Botanical Name	Altitudinal Limit Feet	Altitudinal Limit Meter
1.	Phaphar	Fagopyrum tataricum	13000	3900
2.	Rice		13000	3900
3.	Wheat		13000	3900
4.	Barley	Hordium hexastichen	13000	3900
5.	Potato		13000	3900
6.	Cheena	Panicum miliaceum	11000	3300
7.	Ogal	Fagopyrum esculantum	11000	3300
8.	Kedari Chua	Amaranthus candatus	11000	3300
9.	Chua	Amaranthus frumentaceoun	10000	3000
10.	Chamli	A.blitum	10000	3000
11.	Madua	Eleusine corocana	10000	3000
12.	Mungari	Zea may	10000	3000
13.	Kauni	Setaria italica	7000	14-2100
14.	Jhangora	Oplismeous frumentaceous	7000	15-2100
LEGUMES				
15.	Babla	Viciafeva	10000	3000
16.	Lobia	Vigna eating	7000	2100
17.	Moth	Phaseolus aconilifolium	7000	2100
	Mung	P.mungo		
	Sen	P.vulgaris		
	Gahat	Dolichos biflorus		
	Sheni	D.lablad		
	Bhat	Glycine soja		
	Tur	Cajanus indicus		
	Gwar	Cymopsis psoralisides		
	Masur	Ervum less		

TABLE V

Physical parameters for agricultural landuse Planning

Sl. No.	Land forms	Altitude (Meter)	Climatic conditions: Kemp (C)	Climatic conditions: Rainfall/ snowfall (mm)	PH value	Soil Conditions: EC/mm/ hs cm	Soil Conditions: Organic matter %	Nitrogen	Drainage condition order/ Texture Bifurcation ratio	Broad Land-use: Agriculture	Broad Land-use: Forest pastures	Crop gradient Pattern (Photo-synthesis/ Altitude Gr.
										Hyr/111	Hyad/112	
I	Lower Alaknanda Basin	Below 1200	12-20°	Above 2000	7-8	.10 - .12	8-12	Above 40	5 + 6, 35.7 (2-5) well-drained	Rice Madua, Jhangran Barley Wheat, Toria, Potato	Sal Kanju Semal Haldu	+S, C^4
		SUB-TROPICAL CLIMATE										
										Hyr/221	Hyad/212	
II	Middle Alaknanda Basin	1200-2400	10-15°	1050-2000	6-7	.10 - .12	4-8	20-40	3 + 4 4.6 (5-6) well-drained	Jhangora Marsa, Kaunl Madua, Cheena, koto, ogal, wheat, barley, potato,	Chirpine Banj oak Maru oak	+S + CAM S + CAM $+C^4$
		COOL-TEMPERATE CLIMATE										
											Lwt/321	And/312

Table V continued.....

III	Upper Alaknanda Basin Cold	Above 2400	8-10°	250-1050	Less 6	Less .10	Less 4	Less 20	1 + 2 4.4 Above 6	Marsa, ogal, phapar green beans barley, wheat, potato	Silver fir, blue pine	+5+CAK +C^{4} C^{3} + S + CAM
		COLD CLIMATE									$\frac{Lwt}{321}$	
	Summer crop zone	2400-3000	10°	1050	6-4	.05 - .10	2-4	10-20	-	Fafar, ogal chua, barley wheat	deodar	
	Alpine summer crop zone	3000-4000	below 8°	250 partly (snowfall)	Less 4	Less .05	Less 2	Less 10	-	Fafar potato marsa	Alpine pasture	
											$\frac{TEW}{424}$ (fig)	
	Glacial zone	4200-4800 above	below 0 C^{o} below freezing point	snowfall								

FIVE

LAND CAPABILITY CLASSIFICATION AND LANDUSE PLANNING

LAND CAPABILITY CLASSIFICATION

The existing land use pattern of the region which is evolved from long experiences of the people, has been primarily considered to classify the land for rational land use planning. The standard scheme of the land capability classes has been slightly revised to suit the Himalayan region. To assess the land capability, the parameters of soil genesis, aspect, slope, relief and land system were analysed after giving them due weightage. Land evaluation factor has been worked out as formulated here (Table VI)

$$\text{Land Evaluation Factor} = \frac{(\text{Soil Index})\ (\text{Physiographic Index}) \times \text{aspect}}{100}$$

The land evaluation ratio factor as derived from this method varies from 20 to 90 per cent as applied for classifying the land under land capability classes for rational land use planning (Table VI).

S. No.	Land Capability classes	Description
1.	Class I $\frac{I_{c1}}{(>90)}$	Well protected and high potential for sercitization and chloritezation and irrigated terraced agricultural land. The land evaluation ration is 98 which occurs under the category 90 to formulate $\frac{I_{c1}}{(>90)}$
2.	Class II $\frac{II_{c2}}{(70\text{-}90)}$	Unirrigated agricultural land which is moderately potential for agricultural development with a restricted range of crops needing intensive management. Land evaluation ratio is 71.31 ranging under the category of (70-90)

to formulate $II \frac{c2}{(70-90)}$

3. Class III $III \frac{c2}{(40-70)}$ — Land potential for productive forest development, rationally suited to commercial timber exploitation. Land evaluation ratio is 44.35 which occurs under the category of (40-70) to formulate Factor: $III \frac{c2}{(40-70)}$

4. Class IV $IV \frac{c2.3}{(30-40)}$ — Land suited to forest conservation with little or no mineral, agricultural or forest development potential. Land evaluation ratio is 31.30 which occurs under the category of (30-40) to formulate Factor: $IV \frac{c2.3}{(30-40)}$

5. Class V $V \frac{c3.4}{(20-30)}$ — Land needing an ecological change in its uses, highly suited to pastures and orchards. The land evaluation ratio is 27.96 which occurs under the category of (20-30) to formulate the Factor: $V \frac{c3.4}{(20-30)}$

6. Class VI $VI \frac{c4.5}{(10-20)}$ — Land highly suitable for managed postures. Its land evaluation ratio is 17.60 which occurs under the category of 10-20 to formulate Factor: $VI \frac{c4.5}{(10-20)}$

7. Class VII $VII \frac{c6}{10}$ — Not suitable and easily acessible for biotic operations. The land evaluation Factor VII_{c6}, i.e. nil for agricultural forest and pasture developments.

8. Class VIII $VIII_{c1,2,3,4,5}$ — Miscellaneous land suitable for biotic operations for cultural developments. $VIII_{c1,2,3,4,5}$

Landuse Planning

Landuse planning aims at a much better adjustment of agriculture to the physical environment for optimum utilization and conservation of natural resources. No plan may be final or ever lasting as the cultivation is a dynamic undertaking and may never be static. The factors taken under consideration are related to the land capability, pressure of population and requirements of the people. The land evaluation factor has also been worked out to rationalise the real claim of land. In the valley the land evaluation factor for agriculture ranges from $\mathrm{I}_{\frac{c1}{100}}$ to $\mathrm{II}_{\frac{c2}{70}}$ for forest from $\mathrm{III}_{\frac{c2}{40-70}}$ to $\mathrm{IV}_{\frac{c2.3}{30-40}}$ for pasture land $\mathrm{V}_{\frac{c3.4}{20-30}}$ for managed pasture $\mathrm{VI}_{\frac{c4.5}{10-20}}$ for open grazing $\mathrm{VII}_{\frac{c4.5}{<10}}$ and for other purposes $\mathrm{VIII}_{c1,2,3,4,5}$ (Table-VI).

Class I

The land is suitable for cultivation without special practices. The land evaluation factor of this kind of land is $\mathrm{I}_{\frac{c1}{100}}$ which indicates that the land is well protected and high potential for agriculture. Sericitization and chloritization processes have formed rich soils with all properties. The land is generally irrigated and double cropped. The slope ranges fron 5^{0} to 7^{0} and soil depth is normally found 20 cm.

In addition to all these properties, the land requires certain measures to have better yield. It should be properly terraced and levelled and provided with proper water disposal system so as to protect against gulley erosion. It may also require replenishment of nutrients, provision of irrigation and introduction varieties suited for soil structure, control of pests and diseases. The characteristics of Class I land in the Himalayan terrain should generally of stable surface i.e. river terraces, flood patches, well protected, repose slopes, talus cones which are geneally stabilised and irrgable. The soils usually are deep, fine textured, having very few out-crops and receiving sufficient sunshine to mature arable crops. Thus the land under Class I is properly diagnosed for producing agricultural crops without intensive land management.

Class II

The land is physiographically located on the similar situation as Class I on flat lower hill tops and slide slopes. The land evaluation factor of this kind of land is $\mathrm{II}_{\frac{c2}{70-90}}$ which is moderately potential for agriculture. The slope varies from 7^{0} to 15^{0} and the land comprises of moderate soil depth, erodible soils and coarse fragnents with poor sunshine. Thus this class of land requires intensive land management. However, unirrigability will check the progress of yield. Hence, means of irrigation should also be provided to achieve optimum use of land.

Class III

The land is highly potential for dense forests. The land evaluation factor of this kind of land is $III_{\frac{c2}{40-70}}$ which is rationally suitable to raise goods forests. All existing forests having good soil depth with rockyness and stoneyness. The slope of the land generally varies from 25^{0} to 30^{0} which are manageable. About 198900 hectares of land are available in this category. About hectares of land may also be manageable for forest development.

Class IV

The land, which comprises of uncultivated repose slope, dissected and gullied flood plains and terraces with poor vegetable is suitable for forest development with intensive forest conservation measures. The land evaluation factor of this kind of land records $IV_{\frac{c2.3}{30-40}}$ which needs certain remedial measures as discussed in Table VI. The land where slope prevails over 30^{0} and soil cover is moderate to deep, afforestation may be done with bench terracing. Diversion ditches and melching devices should be adopted to prevent severe erosion. Scrubs (Agave sisalana) may be planted on inter-terraced boundaries.

Class V

The land needs an ecological change in its uses. The land comprises of slope varying from 7^{0} to 15^{0} with shallow cover of soil, which is not suitable for agriculture. Therefore, it is proposed to have a new class in the Himalayan terrain for this land which requires slight changes in its uses in the following manner.

(i) The land which ecologically claims for a particular use, is used at present for the same.

(ii) The land where present landuse is causing detriment to its ecological conditions.

The changes in landuse may be proposed from pasture land to terraced cultivation or forest plantation and from forest to pasture land or terraced cultivation. The land evaluation factor of this kind of land is $V_{\frac{c3.4}{20-30}}$ which is ideally suitable for pasture land or orchards but watershed management programmes are needed to achieve the proposed landuse.

Class VI

The land, consisting of shallow soil cover on newly formed talus cones, valley side slopes and hill side newly born repose slopes, is suitable for scrubs, scrubs and grasses provided the cattles are checked for open grazing. The newly formed land forms are too week to protect them from the adverse effects of mass wasting processes. The land evaluation factor is $VI_{\frac{c4.5}{10-20}}$ which needs soil and water conservation measures to treat this land for managed pastures or grazing yards.

Class VII

The land has shallow patches of soils with rocky and stony characters. Slope varies from steep to very steep. The land evaluation factor is lowest i.e. $VIII_{c1,2,3,4,5}$ which is not feasible to put in under any kind of landuse except for military purposes and tourism. The measures for soil conservation may not be feasible to operate therein.

The treatments of the recommended landuse have been suggested according to the land capacity classes and their evaluation factors as stated in the Table-VI and VII).

Table VI
Parameters determining the landuse capability classes

S. No.	Rating if climatic conditions	Soil Factors: Tex	Depth	DC	Om	Erosi.	Soil index no.	Physiographic Factors: Relief	Slope	Aspect	Physiographic Land Evaluation Index No.	Land Evaluation Factor	Existing land use	Land use Capability classes (✓- eva)
1.	C_1	17.13	20.00	17.13	15.00	40.00	115	50	36.6 x 1		86.6	98	Irrigable cultivated terraced land	I_{C1} (¯90)
2.	C_2	20.00	30.00	25.00	15.00	36.6	126.6	30	26.6 x 1		56.60	71.31	Cultivated terraced land unirrigated	II_{c2} (70-90)
3.	C_2	15.00	16.42	23.57	13.57	30.00	98.56	28.5	17.1 x 1		45.6	44.35	Dense Forest	III_{c2} (40-70)
4.	$C_{2,3}$	16.87	15.00	22.50	13.75	23.75	91.87	22.7	11.4 x 1		34.1	31.30	Sparse forest	$IV_{c2,3}$ (30-40)
5.	$C_{3,4}$	18.00	10.00	24.00	18.00	23.20	93.2	20.00	10.00 x 1		30.00	27.96	Pasture/ scrub land	$V_{c3,4}$ (20-30)
6.	$C_{4,5}$	16.00	10.00	24.00	18.00	20.00	88.00	10.00	10.00 x 1		20.00	17.60	Barren Land	$VI_{c4,5}$ (10-20)
7.	C_6	-	-	-	-	-	-	-	-	-	-	-	Land under glacier and snow	VII_{c6} (<10)
8.	$C_{1,2,3,4,5}$	-	-	-	-	-	-	-	-	-	-	-	Miscellaneous land	$VIII_{c1,2,3,4,5}$

Table VII

Treatments determined according to -eva Factor of the land capability for land use planning

Slope	1	2	3	4	5	6	
Soil Depth	Gentle sloping 7°	Moderate sloping $7^{o}-15^{o}$	Strongly sloping $15^{o}-20^{o}$	Very strongly $20^{o}-25^{o}$	Steep $25^{o}-30^{o}$	Very steep 30°	Altitude (m)
Deep (D) (90CM)	$\frac{I_{C1}}{100}$	$\frac{I_{C1}}{90}$	$\frac{II_{C2}}{90}$	$\frac{II_{C2}}{70}$	$\frac{III_{C2}}{40-70}$	$\frac{IV_{C2,3}}{30-40}$	1000
Moderately deep (Md) (50-90 cm)	$\frac{I_{C1}}{90}$	$\frac{I_{C2}}{90}$	$\frac{II_{C2}}{70}$	$\frac{III_{C2}}{40-70}$	$\frac{IV_{C3,4}}{30-40}$	$\frac{V_{C4,5}}{10-20}$	1000-2000
Shallow (S) (20-50)	$\frac{II_{C2}}{90}$	$\frac{II_{C2}}{90}$	$\frac{III_{C2}}{40-70}$	$\frac{V_{3,4}}{30}$	$\frac{V_{C3,4}}{20-30}$	$\frac{V_{C4,5}}{10-20}$	2000-3000
Very shallow	$\frac{III_{C2}}{40-70}$	$\frac{III_{C2}}{40-70}$	$\frac{VI_{C4,5}}{10-20}$	$\frac{V_{C3,4}}{20-30}$	$\frac{V_{C3,4}}{20-30}$	$\frac{V_{C3,4}}{20-30}$	3000-3600
(20 cm) Extremely shallow (10 cm)		Alpine pastures		$\frac{VI}{<20}$	$\frac{VI}{<20}$	$\frac{VI}{<20}$	
		Glaciers Tourism		SN	SN	SN	4000-5000 and above

SIX

ECOLOGICAL PROBLEMS THEIR TREATMENTS FOR AGRICULTURAL DEVELOPMENT

The Alaknanda basin consists of very high mountains and steep, precipitious slopes. Throughout the area, marginal land farmers have attempted to squeeze out a living from lands not suited for crop production and have not only ruined the entire landscape in the process, but have created hazards for floods and sedimentation in the areas down below. These steep mountaneous areas have their own characteristics response to erosion. The terrain is heavily subjected to severe erosion. Avalanches, landslides, land slips etc. are fairly common. High intensity runoff of bare land surface would not usually permit any trace of soil on these hills. However, there are a few redeeming features which make the slopes sufficiently stable to allow some kind of landuse. Poor impact of raindrops, coarse skeletal soils and fast vegetational recovery are some of the conditions that favour soil protection. The pressure of population except in areas where soil is easily workable, is too little. Soil working is restricted to hand digging. Holdings are small, usually confined near the habitation. Poor accessibility of the area would not provide for major technological changes in the near future. On the basis of these assumptions it may not be necessary to think of major landuse changes. The strategy for watershed control would mostly confine to improving practices within major landuse regions. The survey aspect should therefore aim at locating areas whose potential is not being fully utilized. For example, if marginal land is suited for agriculture, it should ecologically be treated through reducing its pressure of overgrazing and developing plantation agriculture on it.

Thus from deep investigations during field studies certain ecological problems were diagnosed and their treatments were also worked out through participant observation techniques as discussed below:

ECOLOGICAL PROBLEMS

Ecological problems dealing with environmental degradation in the Alaknanda basin seems to be a function of geocatastrophic morphogenesis causing a severe depletion of land resources like glacial, glacio fluvial and inter montane valley land forms, minerals, soils, water and forests as a consequence of the excess of human interferences like unscientific quarrying and blasting for road and dam construction, over grazing, over lopping over fallowing, over ploughing and deforestation.

These prevailing unrationalized human interferences cause major landslides, rockslips and debris slides. It is also observed that the above geographical factors accelerate their functions where slates, and phyllites have been faulted crushed, fractured and powdered due to the severe effect of the thrust zones of the Garhwal Himalayas, trending perpendicular to the Alaknanda from NW to SE.

The present study aims to diagnose the nature of land resource depletion and pin point their distribution and then treat them with ecological measures for the ecoderelament of the basin.

Generally land slides, in the valley occur, when hill-side or valley side slope fail due to the certain geological climatic and biotic factors. The said factors have been analysed with the remote sensing techniques of land-sat imageries and serial photographs and their results were confirmed by field observation.

Geoidentical Conditions for Landsliding

Tectonically the Alaknanda valley comprises of three separable major litho-stratigraphical units known as (i) The Dudhatoli (ii) the Garhwal group (iii) The Central Crystalline group (Kumar and Aggarwal 1975). The zone between Deoprayag to Koteshwar consists of moderately metamorphosed rocks particularly phyllites and quartzites forming Dudhatoli Group, which has been separated by the North-Almora-Thrust trending almost in WNW-ESE direction. Further the rocks of the Garhwal Group as exposed upto Vishnuprayag are primarily quartzites, slates, schist and crystalline calcium carbonates occasionally intruded by metavolcanics. The northern part of the valley ranging from Vishnuprayag to the village Mana, consists of schist, gneiss and granite of the Central Crystalline Group which further continues to the village Niti in the north east direction along the Dhauliganga.

Keeping a view point of drainage of the valley, it has been observed that the folded, faulted and fractured terrain is wholly occupied by perennial and non-perennial ground and underground torrents, streams and springs of the main tributaries of the Alaknanda like the Dhauliganga, the Nandakini the Pindar and the Mandakini forming subparallel, dendritic and underground drainage patterns (P.B. Saxena, 1977). The valley is also enriched with hidden tressure of waters of underground seepage, spring and stream as exposed in the form of Patalganga near Nandprayag. Such a complicated drainage network of the valley plays a dominant role to cause landsliding.

The climatic conditions of the valley generally are governed by the nature and degree of altitude. The snowfall delimits the area above 3000 metres and rainfall starts below it. The record of annual mean snow-fall and rainfall in valley is 38 cm and 162 cm respectively. The valley consists of two unique sub-temperate and temperate zones (ranging from 1200 to 2400 metres sub-temperate and above 2400 metres temperate). Thus the sky, which remains richly humid throughout the year and gets rocks moistened causes chemical and mechanical weathering under operation day and night.

Edaphically the valley is least resistant, with the result the soil erosion occurs very severely. The loose strata of micaceous schist, slate, chlorite schist are light green in colour. They have fine grained bands of phyllites schist, quartzite and crystalline, calcium-carbonate. It is generally observed that the chlorite schists are very susceptible to weathering and finally to soil formation. Chlorite schists are mostly available in the Garhwal Group, therefore the wide spread terraces have been developed in the localities of Augustmuni, Gaucher, Maletha, Nagrasu and Srinagar. In high hills of the valley, the soils are dark greyish, moist, sandy loam, subangular, blocky friable slightly sticky. They have 40 per cent to 60 per cent fragments of quartzite and granite. They are susceptible for block disintegration and other processes of mass wasting. In the low mountains, the soils are generally loamy skeletal, mixed hyperthermic, Udic ustochrepts. Their grains are subangular, blocky and non-plastic, friable. They have 5 to 10 per cent chert rock fragments causing soil creeping, rock slipping. In the valley terraces, the soils are coarse, sandy, mixed hyperthermic typic underthents. They are yellowish brown, moist loamy and sandy, structureless non-sticky. These soils are also least resistant and are affected by gullying process.

Slope as a Factor of Gravity Force

The slope plays a dominant role to activate the gravity

force for land-sliding, soil creeping and mud flowing. The slope on the steep-rock-face known as free face is affected under the influence of weathering (Penk, 1934). The weathering of free face causes debris (scree) to accumulate as talus-cones at the base of the slope. The talus initially produces constant slope at the base of free face to form repose slope as observed along the Almora-Thrust and Main Central Thrust. The highly crushed phyllites crumpled and powdered talcose rocks as found along the Thrust Zone of Garhwal Group, consist of repose slopes near the village Belakuchi. When the crushed phyllites and powdered talcose rocks are converted into clay, may act as lubricant during the rainly season are resulted in (Saxena, P.B., 1980) as generally observed in the Garhwal Group.

Landslides also occur, when the base of talus (forming the toe of repose slope) exist beside the river valley and it is eroded by the undercutting action (Nitya Nand and Prasad, 1972) of river water during the flood period as recorded in the history of the village Belakuchi which was slid down in the year 1970.

Ecological Imbalances

The land resource of the valley like forests, water, mineral, soils and physical landforms have been deteriorated frequently due to excessive erosion and frequent occurrence of mass wasting as a consequence of deforestation, overgrazing, overlopping, overfallowing, overploughing, unscientific mining and quarrying and overblasting for construction of highways and high dams and the neglect of the conservation measures. It has been worked out that the debris slides and rock slips generally occur near Reni, Lata, and Helang villages in the Dhauliganga on account of the removal of vegetal cover for agricultural purposes. It is also investigated that the proper intensity of dynamites, used for blasting the hills or hillocks for road and dam constructions has not been considered scientifically, with the result the remote hills are cracked and slid down during the rainy season as observed in the villages of Bhapkund, Bhagyur, Pandukeshwar and Hanumanchatti.

The seasonal movement of floating livestock loosens the soil cover of the repose slopes. A few of them are made of debris of old landside as observed near the localities of Chhinks and Chameli where graziers keep their winter camps. These graziers generally make harm to the forest by cutting main trunks and branches for fodder. The overlapping is generally taken into illegal practices by the graziers in the whole of the Garhwal. They also graze their cattle on newly developed repose slopes to loosen the upper layer of soil cover, consequently the slope is weakened and rockslips occur. Besides these, trees are also cut mercilessly by the villagers to satisfy their domestic needs.

The soil cover also weakens when the peasants leave their ploughed land as fallow due to some economic problems. Sometimes their overploughing causes soil creeping and mud flowing. Such excessive human interference in the nature causes severe land depletion by the landslides.

Land Depletion and Landslide Intensity Zones

In the present study the remote sensing techniques of land sat imageries and aerial photographs have been used to locate the landslide scars and measure their area to evaluate the nature and intensity of landslides. The primary data for landslide frequently, density and land damaged by landslides were collected from 40 sample villages (Fig.5). In the Garhwal Group, the depletion of agricultural and forest land caused by the landslides, ranges from 1.0 to 1.5 hectares per landslide. The density of landslides per Km^2 varies from 7 to 10 as observed in the villages of Maithana, Chamoli, Pipalkoti, Pakhi and Helang etc. (Fig.6). In the Central Crystalline Group, the depletion of agricultural and forest land ranges from 0.5 to 1.0 hectare per landslide. The density of landslide per Km^2 varies from 3 to 4 in the villages of Pandukeshwar, Hanumanchatti, Dhak, Tapoban, Reni, Jelam and Malari etc. (Fig.5). In the Dudhatoli Group, the agricultural and forest area damaged by loandslide is less than 0.5 hectare per landside. The density of landslides per Km^2 is lowest i.e., ranging from 1 to 3 in the villages of Kaliasor, Rudraprayag, Ratura, Ghultir and Langasu (Fig.5 & 6).

After studying the nature and degree of slope, intensity of landslides (frequently of landslide per annum) density of landslides (number of landslides per Km^2) of the sample villages and land depletion (depletion of agricultural and forest land in hectares per landslide per sample village). With the help of scatter diagram and remote sensing techniques the following landsliding intensity zones (P.B. Saxena 1980)[1] have been made.

1. The Vishnu-Nand-Garhwal High Intensity Zone.
2. The Manna-Vishnu-Malari Central Crystalline Medium Intensity Zone.
3. The Deo-Karan-Nand-Dudhatoli Low Intensity Zone.

ECOLOGICAL TREATMENTS

It is generally observed that the low capable land of

rainfed areas of the valley is being deficient in both organic and inorganic properties. The organic properties (humous) are exhausted by using land without any special practices and inorganic properties are depleted by leaving the land bare-fallow after ploughing two or three times for maize and rice. In fact the efficiency of soil is being reduced by haphazard use of the

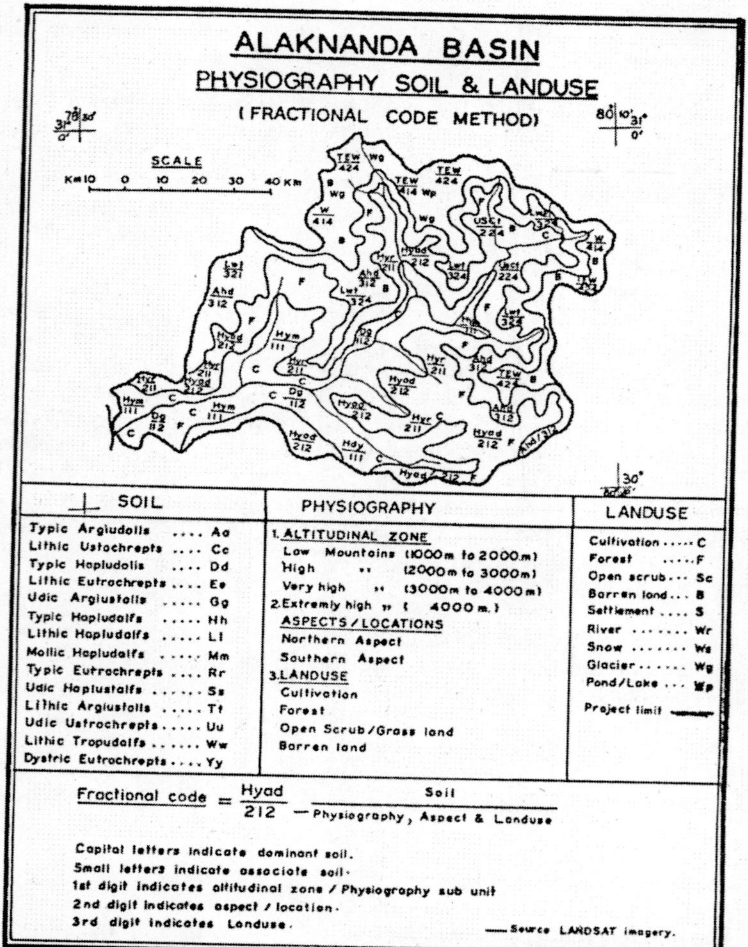

Fig. 5: Alaknanda Basin - Phisiography, Soil and Landuse

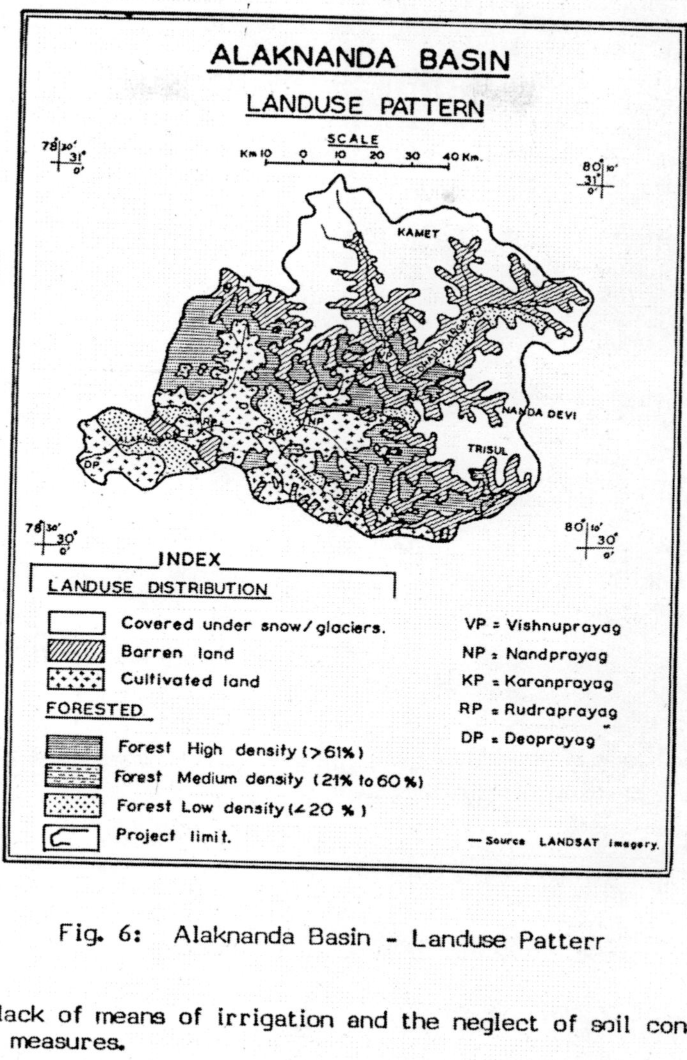

Fig. 6: Alaknanda Basin - Landuse Patterr

land, lack of means of irrigation and the neglect of soil conservation measures.

To protect the valley from the Geographical hazards the following preventive measures are needed to be introduced therein.

1. High power explosives should not be used for road construction.

2. Underground drainage system should be adopted for small channels and streamlets. There should be small drainage channels on road side and water collected from such drainage should be joined to the underground system of drainage to minimise the road side erosion.

3. Stout embankments should be constructed at the toe of the slopes and on the upper side of the roads.

4. To protect roads from erosion, high embankments knitted with iron should be constructed.

5. Large scale deforestation, overlapping and overgrazing and barefallowing should be checked.

6. Unscientific farming on the valley slopes should also be prevented.

7. Afforestation and fruit gardens should be encouraged on the valley sides and spurs.

8. The dry and parennial streams flowing across the roads should be allowed to drain through the cement pipes or iron pipes.

9. The long creepers and Agave sislana should be grown on the bunds of the spur-terraces to protect them from erosion and landsliding.

10. The conservation measures for watershed management should also be brought into practice.

The above mentioned conservation measures are immediately needed to protect the Vishnu-Nand-Garhwal High Intensity Zone which is being depleted and ruined frequently.

To check the soil and water depletion, the following ecological measures are also needed to be implemented.

(i) The soil and water loss, caused by gully formation and fast flow of seasonal and perennial rivers, must be checked and conserved by scientific measures.

(ii) To control the landslides which occur on account of overgrazing, over-lapping, over-blasting, unscientific, quarrying, deforestation, bare-fallowing, fast flowing of torrential runoff, fast infiltration of the drainage, water-shed management projects should be carried out.

(iii) Lift irrigation devices should also be encouraged and developed.

(iv) The chemical and mechanical analysis of the soil survey should also be conducted to treat the soils properly.

(v) Gardening i.e., horticulture should be given priority to control the gully-formation and soil erosion.

(vi) Wood-craft, toy-making, doll-making knitting of wool garments and carpet cottage industries must be given priority.

(vii) Fruit-canning, jelly and jam industries should also be encouraged.

(viii) Technical training centres should also be opened in the rural areas to produce technical hands for agricultural purposes.

(ix) Cultivation of agresislans should be given priority to control the soil and water loss by planting on the field bunds. Agave plantation is very suitable to protect the agricultural fields from graziers and to produce the fine fibres for rope production.

(x) Forest plantation should also be encouraged on the barren land to check the gully-formation.

(xi) Mini-generators should be installed to harness power potential for agricultural and industrial development in the valley.

Measures for Paddy Cultivation

Though the paddy is an important crop in the Alaknanda basin even its yield is very low due to some localised factors which may be treated if following measures are adopted in the system of paddy cultivation:

Application of fertilisers

The paddy crop should be fed with 100 kilograms of nitrogen, 40 kilograms of phosphoric acid and 40 kilograms of potash. All these items should be supplied to the plants in the form of urea, single superphosphate (containing gypsum) and muriate of potash. Before giving the final puddling and just before transplantation, half of nitrogen (50 Kg) and full dose of phosphoric acid and potash (40 Kg) should be supplied for early seeding of the plants.

Control of pests and diseases

To make the crop free of pests and diseases regular preventive measures should be followed. Many posts breed in the wild grass that come up in the field during the off season. Hence regular weeding is essential during the growth of plants. The areas where grass hoppers appear in large numbers every year, the bunds of the fields should be scraped upto 8 cm during land preparation. The plants should be very carefully watched right from the nursery stage. Look into the egg masses laid by the insects like, stem borer, swarming caterpillar and hispa. These eggs should be collected and destroy them as they appear.

In spite of the preventive measures, pests may still appear. In that case the following measures should be taken as pests appar on the leaves and stems of the plants.

During the nursery period about 10 days after germination, the ratio of .04 per cent methyl parathion or .03 per cent Rogor or .05 per cent Lindane should be sprayed after 15 days. After transplantation, crop should be treated with 5 per cent BHc @ 25 Kg/ha in case jessids or gur di bugs or catterpillars or grasshoppars are observed to be damaging the crop.

Disease of Paddy

There has been no serious disease problems in paddy crop in this area. However, it is essential to keep a watch on paddy diseases such as blast, halminthosporium leaf spot and bacterial blight. These diseases are favoured by high humidity. In the years of heavy rains all these diseases or any of them could become a problem. Blast has already been observed in some fields in Delhi and Haryana.

The following steps may be helpful

Treat the seed with Agrosan G.N. or Caresan @ 2.5 grams per kilogram of seed against any external seed borne infection. In case blast or helminthosporium leaf spot is observed and the rainy and humid weather persists, spray the crop with any copper fungicides containing copper oxychloride using concentration of 2 to 3 g per litre of water. The spray may be repeated after every 7-10 days depending on weather conditions and intensity of infection.

Apply balanced fertilizers, because high nitrogen doses alone make plants susceptible to diseases particularly to bacterial blight.

Harvesting, Drying and Storing

Harvest the crop 30-35 days after flowering when the grains are fully formed and the moisture contants of the crop ranges from 20-25 per cent. The harvested crop should be immediately dried and threshed (preferably by a mechanical threshder) to avoid grain losses.

In court-yard grains should be occasionally turned and covred in heaps during the nights hours to avoid dew deposits. For safe storage of grains as seed, moisture content should be in the range of 12 to 14 per cent. The storage should be in air tight container.

The milling should be done preferably where rubber rollers are used. Such rollers reduce friction and thereby reduce the breakage losses. Harvesting at high moisture levels followed by drying improves the milling quality of paddy.

Measures for Cultivation of Toria and Rai

The cultivation of Toria and Rai is highly practised in the Alaknanda basin but its yield is very low. The yield may be increased if following measures are taken in the system of cultivation.

Application of fertilizers

The fertilizers, N.P.K. should be used in soil from 6 to 10 cm below the seed at sowing. Their doses have been recommended as given below:

Nitrogen - 20-40 kg/hect.
Phosphorus - 30-40 kg/hect.
Potash - 30-40 kg/hect.

Sowing time

Toria should be sown from third week of August to the end of September. The optimum time for sowing is the middle of September. Rai however may be delayed for a week.

Seed rate

Seeds of toria and rai should be sown at the rate of 3-4

kg/ha.

Spacing

The distance from row to row should be 45 cm and plant to plant 15-20 cm.

Irrigation

At the time of flowering irrigation is essential if irrigation facilities are available.

Weeding and hoeing

Weeding and hoeing should be done when the crop is about 20-30 days old.

Harvesting

The crop should be harvested when the leaves and pods become yellowish otherwise there is danger of shattering.

Yield

A good crop of Toria will yield on averages 600 to 800 kg/ha and that Rai from 700 to 900 kg/ha.

Pests and diseases

To control the attack of Bugrada crucifurm, 10 per cent BHC @ 0-15 kg/ha should be used after one week of germination. For aphid control, the Rogor (01-0.21 per cent) should be sprayed whenever aphids appear.

Seeds should be treated with Thiram at the rate of 2 cm per kg of seed. Mustard is also sometimes attacked by Alternaria blight disease which produces dark brown spots on leaves, stems and pods. Normally no control measures are needed. But if the attack is serious, any fungicides containing ZINEB at the rate of 2000 gm in 1000 litres of water per hectare should be sprayed.

Citrus Farming

Sweet orange, mandarin, sweet lime, kagzi lime and grape fruit come under the group of citrus fruits. The favourable aspects with regard to their cultivation are discussed below:

CLIMATIC CONDITIONS

Sweet organge and grape fruits

The sweet orange and grape fruits thrive better in sub-tropical dry climate having well defined summer and winter seasons and low rainfall areas. Sweet oranges if grown in the region of heavy rainfall and high humidity to not ripen properly with the result that they become insipid in taste and the colour of the rind also remains green.

Mandarin

Mandarins prefer more humid and tropical summer climate, warm winters and higher rainfall. Extreme cold and extreme heat are not desirable.

Lime and lemons

Limes and lemons are tropical in their climatic requirements. Situations which are warm, moderately moist and free from strong winds are immediately suited. They are most susceptible to frost.

Soil

The ideal soil is loam or sandy loam without any impervious sub-soil. The most satisfactory pH for Citrus ranges between 6.0 to 6.5. Saline or alkaline soils and soils having lime nodules or water logged conditions should always be avoided. The soil must be well drained and water table should not be high. A fluctuating water table which comes about 5 feet in some season is worst than the perpetually high water table. Such condition ideally prevail in the upper Middle Alaknanda Basin.

Suitable varieties

Sweet orange	Mosumbi and pine apple.
Grape fruit	Dupcan, Foster, Thompson and Marsh
Mandarin	Kinnow
Lime	Kagzi Kalan
Lemon	Nepali ablong

Selection of the plants

The plants should be from the pedigreen trees and of known variety. The plants should be free from virus, insect and fungus infections. But union should be smooth and at least 20-25 cm above the ground level.

Propagation and rootstock

Rootstocks are raised from freshly extracted seeds and sown in the month of August-September. If the sowing is delayed the cold weather sets in and the germination is poor.

'T' budding is the commonest method of propagation for sweet orange, mandarin and grapefruit, etc. Budding is generally done either in spring or in September-October. The rootstocks commonly used are Jambhiri, Rough lemon and Karnakhatta.

Planting and care of orchard

The budded plants are ready for transplanting in the field in about a year's time. Citrus is planted in previously dug and filled pits measuring 3/4 x 3/4 x 3/4 metres. These should be filled after mixing the top soil from the pit with 25 kg. of F.Y.M. BHC 50 per cent 200 grams should also be applied in the pit to check white ants. The plants should be set in the field in February or in September-October. Pat earth round the ball thoroughly to avoid air space. Care should be taken to keep the bud union always 25 m above the ground level.

Irrigation

The field should be irrigated just after the planting. Young plants upto the age of five years may be irrigated by ring method, while the other ones should be irrigated by flooding or furrow method. In summer the plants are watered at 10-15 days interval while in winter the irrigation may be given after every three or four weeks. During the rainy season water is rarely needed and arrangement to drain excess water should be made. Over irrigation should always be avoided and the irrigation water should not come in direct contact with the tree trunk. It can be done by earthing up around the trunk.

Hoeing and weeding

Complete absence of tillage in the citrus orchard is better than deep tillage. The feeder roots of the citrus trees are very shallow and are easily destroyed by deep tillage. Keep the basins clean of weeds all the time.

Choice of inter-crops

In choosing the inter crops, avoid those crops which require heavy·irrigation, especially in winter, like berseem etc. Cultivation of inter-crops like beans, cowpeas during hot months, if sufficient water is available, is very beneficial. It would increase the orchard humidity during this period and reduce the fruit crops. Growing of green manure crops like sunhemp or guar during the rainy season would keep the weeds under check and help in the transpirating of excessive water from the field after heavy rains.

Manuring

The manure as suggested below may be added:

Before planting 20 kg or one basket F.Y.M or composit + 1 kg of superphosphate to each pit and mixed thoroughly.

After planting till flowering

Age of the tree	F.Y.M	Kilogram per tree		
		Ammonium sulphate	Super-phosphate	Muriate of potash
1	20	1/4	1/4	1/4
2	25	1/2	1/4	1/4
3	30	1	1/2	1/2
4	40	1 1/2	1	1/4
5	50	2	2	1

Bearing trees

60 kg of FYM + 2 1/2 kg Ammonium sulphate + 2 1/2 kg superphosphate + 1 1/2 muriate of potash.

The above mentioned fertilizers should be split in two equal doses. First before January and the rest after fruit set followed by irrigation. FYM should be supplied in November-December.

Pruning and training

Training to build up a strong frame work begins fron the nursery. All branches that start within a few inches of the union be removed leaving about 1 1/2 to 2 feet of chean straight stem, retaining four or five well spaced branches. All the diseased, injured or crossed branches and dead woods should be removed periodically. Water shoots arising from the base of bud union should always be rippened in the young stage. The pruning wounds should be covered with bordeaux paste, which contains copper sulphate (1 kg) lime ($1^1/2$ kg) and water 15 litres.

Nutrient deficiency and dieback of twigs

Chlorosis of young and old leaves is common. The leaf size is reduced and dieback starts. It may be due to deficiency of zinc alone or combination of manganese, copper and iron etc. A spray of the following ingradients is most useful. In order to secure maximum effect, the application should be done after the major flush of leaves has expanded to at least 2/3rd or normal size.

Zinc sulphate	$2^1/2$ kg
Copper sulphate	$1^1/2$ kg
Manganese sulphate	1 kg
Magnesium sulphate	1 kg
Ferrous sulphate	1 kg
Boron	$^1/2$ kg
Lime	$4^1/2$ kg
Urea	5 kg
Water	450 kg

Preharvest fruit crop

Preharvest fruit crop may be serious problem especially in case of mandarin and sweet oranges. It can be minimized by spraying 10 ppm (1 g in 100 litres of water) 2,4-D and 2,4-5-T in May and August.

Harvesting

Harvesting of fruits starts from August, September (lemon, lime etc.) and December (sweet orange, mandarin and grape fruit) and it requires some special precaution, while picking the fruits, the stem end should be cut close to the fruit without damaging the ring.

Insect, pests and diseases

(1) Citrus leaf minor (Phyllocnistia citrella)

The leaf minot is the larva of a tiny month which lays almost invisible eggs on the leaves and shoots. On hatching the larva bores under the epidermis and feeds on the plants cells by weaving a serpenting tunnel on the young leaves and terminal growing shoots. It can be controlled by spraying the plants with

0.1 per cent metasystox or 0.035 per cent phosphmidon or 0.03 per cent Methyle dameton.

(2) Citrus psylla (Diaphornia citri)

Both the adults and nymphs of the pest, such the sap from the tender part of the buds, leaves and branches and inject into them intoxic substances. In the event of severe attack, leaves get distorted, become sickly, curled up, wither away and ultimately fall off from the plants, resulting in complete defoliation of the plants. It can be satisfactorily checked by spraying the plants with 0.5 per cent malathion or 0.02 per cent endrin.

White fly

White fly nymphs and pupae are found on the under surface of leaves. They cause serious damage by sucking the sap. Spray 0.1 per cent parathion.

Trunk borer (Inderbela sp.)

The caterpillars bore into the wood and damage them by eating the bark. The most common control measure for Inderela caterpillars consists of injecting the bales with fumigants like kerosene, petrol and carbon disulphide. Injection of 0.013 per cent dichlorves or 0.05 per cent trichlorfon or 0.05 per cent endosuffan or 10 per cent kerosene soap emulsion are also effective and cheap.

Citrus canker (X anthomopas citri)

It is the most commonly prevalent disease of kagzi lime during the rains. It affects the leaves, twigs thorns and fruits. The lessions appear as small yellowish spot which enlarge to a diameter of 3 or 4 mm and become raised and rough or spongy and turn brown.

The infection can be largely prevented by removing the infector spraying 3-4 times in a season with antibiotic the streptocycline (Hindustan Antibiotics) at the rate of 1 g in 45 litres of water or intermittantly spray of a suspension 'Neem Gake' at 1 kg in twenty litres of water.

Guava

Guava (Psidium quajaqva L.) a native of tropical America is grown in almost all parts of India in tropical and subtropical regions like U.P., Rajasthan, Haryana, Punjab. It is very hardy, long lived and needs little attention. It is excellent source of vitamin C and essential minerals like calcium and phosphorus.

Climate and soil

It requires dry moderate winter and summer seasons with an annual rainfall not exceeding 100 cm restricted to four months from June to September. Guava is highly susceptible to frost particularly in young stage. It succeeds in every type of soil (poorly alkaline to poorly drained) from high sandy loam to clay soils pH ranges from 4.5 to 7.5.

Varieties

In north India there are well known varieties. They are Allahabad Safeda Red. Fleshed Chittidar and Lucknow 49 fruits of Safeda are round smooth skinned white fleshed sweet and excellent in flavour. Lucknow 49 is a variety bearing large round meaty white flesh fruit of excellent quality. The surface of fruit is not so smooth. There is also a seedless variety, which is shy bearer and fruit are of irregular shape.

Propagation

The guava is usually propagated from seed but it is not desired, because the seedlings of guava are not true to type. Hence vegetative propagation should be practised as mentioned below.

(1) Stooling

The method may be used for quick multiplication of uniform rooted plants. The layered plants grown in situ for 2-3 years are headed back near the ground in March. A ring of bark about 2 cm wide is removed from the base of each shoots just after the onset of monsoon followed by earthing of the shoots. In September rooted plants are separated.

(2) Air layering

A ring of bark about 2.5 cm long is removed below the bark from the portion of the selected shoots. It is covered with moist moss and wrapped with alkathene.

Source: Division of Horticulture and Fruit Technology. Indian Agricultural Research Institute, New Delhi-110 012.

Advantages

(I) A plot of 4.5 square meter can yield about 300 rooted plants each year and the same plot may yield the same number rooted plants for several years, and (II) the cost of production is very less.

Planting

The best time for planting in the Alaknanda Valley is from the last week of June to August at a distance of 6-7 meters apart. The plants may also be planted in February during spring.

Cultural practices

The care of young trees consists of regular watering during the dry season, hand digging the basins occasionally and removing water shoots. Three to four good, strong well grown branches, radiating from the trunk at a height of about 90 cm on all the sides should be selected. Young trees should be trained by prunning so as to give them a definite shape. All dead branches should be removed. Young trees should be watered once a week during the hot weather and twice a month during winter.

Young trees should be manured once a year in the beginning of monsoon. A one year old tree may be given about 12-15 kg FYM. This quantity should be increased by 15 kg every year upto the end of fourth year. A big plant (10 years or more) should get about 100 kg of FYM and 2.5 kg of ammonium sulphate, 1.25 kg of potassium sulphate and superphosphate each, should also be added alongwith ammonium sulphate. FYM should be given late in winter fertilizers in two equal split doses in February and July Ist week.

Fruiting

The rainy season fruits are insipid and watery and do not keep well. Rainy season fruiting should be discouraged, so as to get good crop during winter. Winter season fruits are excellent in quality and escape the attack of fruit fly. To discourage the fruiting during rainy season, do not water after taking winter fruit. Digging and manuring should start just before the rains. This will help in giving good winter fruits.

Yield

A seedling trees of 8-10 years and above may bear 400-500 fruits weighing about 80-100 kg and a grafted or a layered tune of the same age may give more yield.

Pests

Fruit fly

It is a serious pest on the rainly season crop. Its control consists in preventive measures only. Collect all the affected fruits and burn them. The insect lay their eggs inside the fruit. The maggots make the fruit unfit for consumption.

Diseases

Guava wilt

It is a fungus disease. Branches of grown up trees wither and die one after another. It is advisable to remove and burn the trees that show the symptoms of the diseases. Avoid water logging and alkaline soils.

Reference

1. P.B. Saxena 1980, A Geographical Study of Landslides : Its Genesis, Intensity of Soil Depletion and Zoning in the Alaknanda Valley (Garhwal Himalaya), **Perspectives in Geomorphology,** H.S. Sharma (ed.) 1980.

SEVEN

A CASE FOR SOIL AND LAND USE SURVEY IN THE PEDOSERICITIC ZONE OF THE AUGUSTMUNI BLOCK BETWEEN GARHWAL AND DUDHATOLI GROUP IN THE LOWER ALAKNANDA BASIN

GEOIDENTITY

The Rudra-Karan Pedo-Sericitic zone of the Garhwal Group in the Lower Alaknanda basin lies in between 30'15' to 30'25'N latitudes and 78'55' to 79'15'E longitudes with the area of 544.46 km^2. The surveyed zone includes the sericitized and chloritized quartzites phyllites and chlorite schists in between Rudraprayag and Karanprayag where rivers Pindar and Mandakini constitute their confluences with the Alaknanda river (ranging from 450 metre to 1800 metres from the sea level)

The zone under investigation was selected because of its ideal conditions for the faster growth of soil genesis (sericitization, chloritization and disintegration of rock). In fact the zone is tectonically influenced by crumpling, thrusting, faulting and bending movements of the North Almora thrust, the Alaknanda thrust and the Main Central thrust which exist around it. Consequently the mechanical weathering (disintegration of rock by stress and strain of the thrusts) might have been activated in great extent during the period of tectonic movements. Under such geomorphic ideal conditions the river Alaknanda and its tributary the Mandakini constitute wide ingrown meanders causitive to broad alluvial terraces of the villages of Gauchar, Nagrasu and Augustmuni etc in the zone.

Geologically the zone consists of quartzite granite, dolomite, phyllite orthoquartzite and chlorite schists of the Garhwal Group. The major part of the Rudraprayag - Tilwara zone is occupied by sericitized quartzites, granites and intrusive of metabasics. Orthoquartzite, grits, shale, slits are found in between Karanprayag and Nagrasu. The bands of dolomite, black slate and phyllite, occurring as a narrow band in between Rudraprayag and Ratura, have been referred as the Calic-zone of Lamen.

Physiographically the Lower Alaknanda Basin consists of fluvial unique land systems of mountains and river valleys. Mountains, which play a dominant role in the existing land use, occupies percentage area of the zone ranging from 1000 to 2000 m high from the sea level and they are classified under (i)Low mountains and (ii) Very low mountains clad with dense vegetation on the northern aspect and scrubs with agricultural strips on the southern aspect. Slope generally varies from 7° to 30°, i.e., gentle to moderate. Though the isolated hills of the water device have steep slope i.e., 30°.

The river valleys are generally narrow. They are characterized by river incision. The ingrown meanders of the valley form alluvial terraces and the intrenched meanders consist of very narrow valleys with escarpments. Talus cones are available where free face slope occurs. The terraces occupy fairly wide area on the valley sides of the rivers. They are generally composed of sandy alluvium, silts, pebbles and gravels as well developed on different levels. The terraces are the grainery of the Alaknanda basin.

The climatic conditions for moderate to rugged terrain of the Lower Alaknanda Basin, slightly differ from place to place due to change in relief and aspect. The degree of diversity in relief contributes to a great extent to the irregularities of climatic elements. The mean annual rainfall occurs from 250 mm to 2500 mm and mean annual temperature ranges from $0^{\circ}C$ to $20^{\circ}C$. The valleys are generally found hot and subhumid during summers and the ridges of mountains clad with thick vegetation are found pleasantly cool. During winters the valley temperature reduces abruptly due to air drainage and mountains are influenced by frost action.

The zone under investigation is well drained by the perennial streams of the river Alaknanda and its tributaries. Generally the tributaries of the Alaknanda river following the direction of thrusts (the Main Central thrust, the Alaknanda thrust and the Almora thrust) trending from NW to SE crossing the river Alaknanda meet at about right angle forming subrectangular and subparallel drainage. The imprints of contorted and underground drainage are recognised in the non-perennial streams in crumpled faulted and thrusted zones of Alaknanda and Almora thrusts. The dendritic network, which is woven by the head streams, exists on the water divides of the zone. Thus the zone is sculptured by the voluminous river like Alaknanda and its tributary the Mandakini forming broad terraces where the valley becomes open and escapments it becomes narrow.

The natural vegetation of the lower Alaknanda Basin consists of broad leaved dense forests. The main species of banj, oak, kharsu, sal are found in the zone. Other miscellaneous broad leaved species as chir, deodar and open scrubs are also found commonly in the zone. The chirpine is dominating species as found in between 1000 m to 1800 m on the north eastern aspect of the mountains. The southwestern aspect of the low mountains are warmer than north-eastern aspect and contain more sunshine and less moisture. Consequently thick vegetal cover is observed on the north-eastern aspects of the mountains and scrubs and narrow strips of agricultural land on the south-eastern aspect of mountain as well as in the valley.

INTERPRETATION OF LAND SYSTEMS AND THEIR CLASSIFICATION

The fluvial land-forms of the Lower Alaknanda Basin consists of mountains and valley structures which are subdivided after using land sat imagries and considering their nature of physiographic elements (relief, slope and aspect) to apprise distinctly soils and landuses in the zone. The mountains and the valley land forms have been discussed in support of landuse programme of the zone.

1. Mountains
2. River valleys

1. Mountains

About 70% of the valley is occupied by mountains. The zone under study is characterised by the abrupt changes in the altitude, slope and in the magnitude of relief variations. Thus the mountains are subdivided into two land systems:

1.1 Very Low Mountain - 1000 m.
1.2 Low Mountain - 1000 - 2000 m.

1.1 Very Low Mountains are generally associated with river valleys. The slope varies from 20% to 30%. The total area of the very low mountains occurs 40% of the total land of the zone. The soil cover is very deep and gravelly. The dominant soil is fine loamy (Table IX). About 29% land is available under agriculture which is highest in the whole of the Alaknanda valley.

1.2 Low Mountains comprise of moderately steep to steep slope varying from 20% to 40%. The soil cover is shallow and gravely. The dominant soil is fine loamy. They occupy 45% area of the total zone. About 60% land of the low mountains in the zone is occupied by forest and rest of 40% is under scrubs and narrow strips of agricultural land. The fringe part of the mountains is badly depleted by severe erosion. Thus 30% land of the low mountains occurs under barrin.

2. River Valleys

The lower Alaknanda Basin consists of very complicated structures of the river valleys as characterised by the incision of the rivers. The total area under river valleys consists of 46.10% of the zone. About 45% of land of the river valleys is occupied by the agriculture. The valley structures comprise of following land system units:

2.1 River terraces.
2.1.1 Alluvial terraces (river built terraces).
2.1.2 Structural terraces (river cut terraces).
2.2 Talus cones.
2.3 Valley escarpments.

2.1.1 River built terraces are known as alluvial terraces which consist of coarse to medium textured alluvial material deposited by the river on the either side or on both the sides. The alluvial terraces in the Lower Alaknanda Basin are broad with gentle to moderate slope as observed in Augustmuni, Gauchar, Nagrasu and Meletha villages of the zone. The alluvial terraces consists of about 15% land of the total area of the zone.

2.1.2 Structural terraces are known as river benches, which are observed on the waxing and wanning slopes of the spurs. They are generally formed by river erosion as observed in Chopta, Mohankhal and Tangani villages. They are semi-annular and rectilinear in shape. The total area of these terraces is 70% of the total land of the zone.

2.2 Talus cones consist of conical forms of unconsolidated debris. The old talus slopes (Repose slopes) are usually cultivated. The upper reaches of talus cones are found dry and occupied by scrubs and bushes. The total area under talus cones consists of 5% of the total land of the zones.

2.3 Valley escarpments are the steep sided ridges and stacks of denuded spurs recorded as isolated relics in valley. The intrenched meander of the river consists of sharpened bluffs and steep sides edges of the isolated relics as observed in the field study.

CLASSIFICATION OF SOILS

The soils have been classified on the bases of above analysis. The soil samples which were selected from representative physiographic sub-units were identified at the highest category level. The alluvial soils are richly marked out in the Lower Alaknanda Basin in patches which have an effect of time and some of them have young soils of minimum profiles. Such soils fall under the order of Entisols. The variability in percentages of clay and of organic carbone is a reflection of stratified alluvium. Entisols are dominance of mineral soils materials and absence of a distinct pedogenic horizons

In the soil profiles as analysed in **talaon** land, one or more pedogenic horizons of alteration or concentration with little accumulation of translocated materials other than carbonate have been marked out. Such type of soil is classified under Inceptisol. Inceptisols have finer soil texture than loamy sand and weatherable minerals and moderate to high capacity of clay fraction to retain cations. They have rock at shallow depth. On recent alluvial deposits either in the form of terraces or in repose slopes the inceptisols are developed.

On the Low and Very low mountains clad with forests, there occurs mollisols in great extent. Mollisols are also recorded in the **talaon** land of the valley. They are combination of very dark brown to black surface horizon that make up more than third of combined thickness of A, B horizons i.e. 25 cm and have high degree of base saturation and relatively deep penetration of organic carbon. The soils of the survey area have been classified according to the comprehensive system of soil classification USDA (soil taxonomy, 1975) upto family and their association. The soils generally fall under three orders:

(i) Entisols
(ii) Inceptisols
(iii) Mollisols

The classification of low categorical level is systematically shown in the Table-VIII.

LANDUSE AND AGRICULTURAL SYSTEM

It is worked out that physiography plays a dominant role in rationalizing the landuse pattern in the hilly terrain. The Alaknanda Basin primarily consists of mountains and river valleys. The river valleys, which have major portion of land under agricultural use and minor portion under open scrubs and barren land, consists of numerous stretches of wide spread alluvial terraces as formed on the meander land of the river. The other important physiographic sub-unit is mountain. Generally mountains ranging from 450 to 1800 m have been dominantly occupied by forests and scrubs. The gentle to steep slopes of the northern aspects are generally covered under dense forest whereas the steep slopes of the southern aspects are found under open forest or barren land. The detailed mapping of the landuse pattern of the Rudra-Karan Tract of the Garhwal-Dudhatoli group of the Lower Alaknanda Basin has been conducted as noted below:

Table VIII
Landuse Pattern

Code No.	Major categories		sub-categories	Area (Km^2)	Percentage	
1.	Cultivated	(I)	Irrigated cropped land	101.00	18.57	
		(II)	Unirrigated Fallow land	150.83	27.53	46.10
2.	Forest	(III)	Dense forest	117.92	21.30	
		(IV)	Medium forest	88.05	16.65	41.20
		(V)	Sparse forest	17.48	3.25	
3.	Scrub land	(VI)	Scrub land	60.92	11.18	
4.	Barren land	(VII)	Barren land	5.60	1.02	
5.	Miscellaneous land	(VIII)	Miscellaneous land	2.66	0.50	
			Total	544.46	100.00	

Of the total land of the sampled area about 251.83 km^2 land is under agriculture i.e. 46.10% of the total area. The area under cropped land consists 60% of the agricultural land which constitutes the highest ratio in the whole Alaknanda Basin. Out of the total cropped land about 4.24% land is irrigated through 'guls'. The total land lying as fallow constitutes 40% of the total agricultural land i.e. 18.57% land of the surveyed area in the lower Alaknanda Basin.

The land under forest constitutes 41.20% of the total area comprising 18.57%, 21.30% and 3.25% land under dense, medium and sparse cover of forests respectively. Area being laid down with thick alluviums and diluviums, consists of highest ratio under dense forest. The land under sparse forest is lowest in the forested area. The area under scrub land is about 60.92 km^2 which constitutes 11.18% of the total land of the surveyed area. The scrub land is mostly available on the steep slopes of the southern aspects of low and very low mountains. The area under barren land comprises of 1.02% of the total land of the study area depicts the lowest ratio to the total surveyed area in the Lower Alaknanda Basin of the total catchment area of the Alaknanda river.

The miscellaneous land which includes the land under settlements, roads and ponds etc. consists of .50% of the total land of the surveyed land. Settlements are generally located beside the agricultural land.

After brooding over the results of the above table, it is worked out that the people mostly earn their livelihood from agriculture and agriculture is the primitive constraints of the topography. About 87% people are engaged in the agricultural activities.

The cropping pattern varies according to different conditions of altitude, climate and soil. Various agronomic cultural practices and tradition have evolved existing terracing, crop rotation and traditional 'gul' irrigation system. The farmers generally harvest both coarse and fine grains. Madua, or koda, Jhangora and Marchha or Chuwa are of coarse grained varities which are mostly harvested from unirrigated land known as uproan land. The paddy and wheat are harvested from the irrigated and rainfed land of the valley but the low land which is irrigated through 'guls' is known as **Talaon** land in the local dialect.

In the Lower Alakananda Basin, important crops (in rabi and kharif seasons) like rice, jhangora, madua, maize and wheat are harvested. Legume crops are generally grown in kharif seasons.

Morphology of agricultural fields

Generally the shape, size and morphology of agricultural fields are coined primarily by the nature of physiography and drainage. The annular and triangular fields are observed where river bends and confluences exist as available in the villages of Maletha, Nagrasu, Paprasu and Dhari. Broad parallel alluvial terraces are found in the localities of Augustmuni, Chandrapur (Kedar valley), gauchar, Nagrasu and Langasu where valley has been broadened and paired alluvial terraces formation took place. These terraces are very wide, thick and long. The annular field pattern exist on the convex slope of the spurs of low and very low mountains. The river benches as observed in the rectilinear shape exist where the river valley is associated with denuded stacks and stumps of the spurs as observed near Karanprayag and Rudraprayag where the limb of anticline also appears in distorted shape. The fields of the talaon land (low land) are mostly irrigated by 'guls' and fields of upraon land (high land) are found rainfed due to lack of means of left irrigation. The change of field morphology affects the way of tilling of fields.

CROP ROTATION

In the Garhwal Himalayan, six crop rotation cycles have been evolved after a long experience. Out of them two crop rotation cycles prevail in the lower Alaknanda Basin as discussed below:

S.No.	Cycle of crop rotation	Code
1.	Cycle of 1 year two crop rotation	$(Cy^1{}_{2CR})$ Lower/Middle Alak.
2.	Cycle of 2 year three crop rotation (Sar System)	$(Cy^2{}_{3CR})$ Lower/Middle Alak.

1. Cycle of 1 year two crop rotation ($Cy^1{}_{2CR}$ Lower/Middle Alak).

Generally this crop rotation prevails in talaon land (low land) which is well irrigated in the Lower Alaknanda Basin. Irrigated terraces on the talaon land are generally extended towards the valley bottom. In such fields two crops are harvested each year without fallow operations. Whenever the fields are not adequately irrigated, Jhangora and wheat crops are harvested.

2. Cycle of 2 year three crop rotation ($Cy^2{}_{3CR}$ - Sar System lower/Middle Alak).

The said crop rotation is practised in the Upraon land where fields are unirrigated (rainfed) and no means of irrigation are available. This cycle of crop rotation follows a particular sar system. In this system farmers divide their fields in two components (sars). In the first year, one half of the field grows rice and the other half grows madua (from April to Sept.) pronounced as 'Satiyara' and 'Kodara' respectively in the local dialect. Both the satiyara and Kodara sar have alternate crops operating fallow devices timely to utilize the maximum gain of carrying capacity of land as noted below:

Sar System

1st Sar Satiyara sar (sar of rice) Crop	Period	2nd sar Kodara sar (sar of madua) Crop
Rice	April-Sept.	Madua
Wheat	Oct.-April	Fallow
Madua	April-Sept.	Rice
Fallow	Oct.-April	Wheat
Rice	April-Sept.	Madua

INTERPRETATION OF LAND SYSTEM, SOILS AND LANDUSE RELATIONSHIP

Each physiographic subunits with the major soils and their association encountered therein are treated herewith (Fig.6).

1. Mountains

Mountains generally occupy 80% area of the valley. The region is characterised by very high relief variations and gentle slope to extremely steep slopes, sometimes leading to escarpments and hanging valleys which throw light on the incision of the perennial rivers and simultaneously up heaval. The magnitude of the relief variation may be understood by the range in elevation from 740 m to 3680 m above. In the area under investigation, mountains have been classified in two subdivisions as given below:

1.1 Very Low Mountains - below 1000m from m.s.l.
1.2 Low Mountains - 1000 to 2000.

The aspect of mountain which have considerable effect on soils and land use has been further subdivided into sub-units.

1. Very Low Mountains Northern Aspect - 1.1
2. Very Low Mountains Southern Aspect - 1.2

1.1 Very Low Mountains

(a) **Northern Aspect Cultivated:** The said physiographic sub-unit is generally found near the valleys with a break of slope. The slope varies from 10% to 30% which are generally occupied by the cultivation. They have been recognised on the photographs by slope pattern, light tone aspect and landuse. The soils are deep and gravelly. The area under cultivation is 21.563 km^2.

Soils - The dominant soils is fine loamy, typic, hapludalfs association with loamy skeletal/fine loamy, dystric Eutrochrepts. These are some inclusions of fine loamy Molli Haplu dalfs.

Landuse - These are under cultivation.

(b) **Northern Aspect Forested:** Such type of land system comprises of moderately to steep land with the slopes varying from 30% to 40%. These are moderately eroded. The soils are generally deep and gravelly. They are recognized on the aerial photographs by their low relief, dark tone, texture, pattern, slope, aspect and landuse. The area under use is 10.063 km^2.

Soils - The dominant soil is loamy skeletal, typic, Hapludalf in association with loamy skeletal/fine loamy typic Argiudolfs.

Landuse - Broad-leaved and their forest.

(c) **Northern Aspect Open Scrub/Grass Lands:** These comprise moderately steep to very steep lands with the slope varying from 40% to 60%. They are recognised on the areal photographs by their relief, gray tone, pattern, aspect and landuse. The soils are generally shallow and gravelly. The area under coverage is 4.687 km^2.

Soils - The dominant soil is loamy skeletal typic, hapludalfs. There are some inclusion of loamy skeletal Dystric Eutrochrepts.

Landuse - They are under open scrub/grass land.

1.2 Southern Aspect Cultivation (Very Low Mountains)

(a) **Southern Aspect Cultivation:** These are the valley side slopes with a break in slope. They comprise moderately steep to steep slope from 30% to 60%. The soils are generally deep and gravelly. Area coverage is 18.365 km^2.

Soils - The dominant soil is loamy skeletal, fine loamy, Udic Haplustalfs in association with fine loamy/loamy skeletal Udic Ustochrepts.

Landuse - They are under cultivation.

(b) **Southern Aspect Forested:** These comprise moderately steep to steep and with the slope varying from 40% to 60%. Erosion occurs moderately. The soils are generally shallow and gravelly. The area under this landuse is 8.125 km^2.

Soils - The dominant soil is loamy skeletal, Udic, Argiustools.

Landuse - Broad-leaved and chir forest.

(c) **Southern Aspects Open Scrub/Grass Lands:** This unit comprises of moderately steep to steep lands with slope varying from 30 to 75%. The crumpled, powdered, and fractured rocks are found near Nandprayag adjacent to the central thrust. The area coverage is 13.125 km^2.

Soils - The dominant soil is loamy skeletal, udic haplustalfs in association with loamy skeletal udic, ustochrepts. There are some inclusion of fine loamy udic, argiustools.

Landuse - The land is used under open scrub/grasses.

1.2 Low Mountains

(a) **Northern Aspect Cultivation:** The cultivated land on northern aspect steep to steep slope varying from 30 to 70%. The soils are moderately eroded. They are recognised on the photographs by their high relief, light tone, pattern, texture, aspect, slope and landuse. The soils are shallow and gravelly. The area coverage is 119.00 km^2.

Soils - The dominant soil is fine loamy, typic, hapludalfs in association with fine loamy/loamy skeletal Dystric Eutrochrepts. They are some inclusion of loamy skeletal, typic eutrochrepts.

Landuse - The land is used under cultivation.

(b) **Northern Aspect Forested:** This unit comprises of moderately steep to steep land with slope varying from 40% and above. Rocks have been found, crushed, powdered and fractured near the central thrust zone. They are recognized on the photographs by their high relief, dark tone, texture, slope aspect and landuse. They are generally shallow and gravelly. The area coverage is 120.25 km^2.

Soils - The dominant soil is loamy skeletal/fine loamy, typic, hapludalfs in association with loamy skeletal, dystric eutrochrepts. There are some inclusion of loamy skeletal, typic, agriudolfs and fine loamy, typic hapludolfs.

Landuse - The land is used under chir forest.

(c) **Northern Aspect Open Scrub/Grass Lands:** On the northern aspect of the low mountains the slope varies from moderately steep to steep varying from 40 to 80%. Severe erosion is generally noticed. The soil cover is shallow and gravelly. The physiographic character is identified on the photographs by high relief grey tone, slope, texture, aspect and landuse. The area coverage is 18.56 km^2.

(a) **Southern Aspect Cultivation:** This physiographic unit comprises of moderately steep to steep slope varying from 25 to 40%. The soil cover on the southern aspect is deep and gravelly. Slight to moderate soil erosion is observed on this aspect. The area coverage is 78.19 km^2.

Soils - The dominant soil is loamy skeletal/fine loamy udic ustochrepts in association with fine loamy/loamy skeletal udic haplustalfs.

Landuse - The land is under cultivation.

(b) **Southern Aspect Forested:** The southern aspect of low mountains under forest comprises of moderately steep to very steep slope varying from 25 to 75%. The soil cover is being gradually eroded. The physiographic character is identified on photographs by high relief, dark tone, texture, pattern, aspect slope and landuse. The soil cover is generally shallow and gravelly. The area coverage is 85.063 km^2.

Soils - The dominant soil is loamy skeletal udic haplustalfs in association with loamy skeletal udic argiustolls. There are some inclusion of fine loamy/loamy skeletal udic ustochrepts and fine loamy lithic haplustafls.

Landuse - The land is used under broad-leaved and conifer forest.

(c) **Southern Aspect-Open scrub/Grass Land:** This physiographic unit consists of moderately slope to steep slope varying from 30 to 70%. The landscape is eroded. The soil cover is shallow and gravelly. The physiographic character is identified on the photographs by high relief, grey tone, slope, aspect and landuse. Area coverage is 26.563 km^2.

Soils - The dominant soil is loamy skeletal udic ustochrepts in the association with loamy skeletal udic haplustalfs. There are some inclusion of loamy skeletal lithic ustochrepts and loamy skeletal lithic argiustolls.

Landuse - The land is used under scrubs and pasture lands.

(d) **Southern Aspect-Barren:** This physiographic unit consists of steep to extremely steep slope varying from 45 to 88%. The rock outcrops are more than 33% of the area. They are identified on the photographs by high relief, light tone, slope aspects, pattern and landuse. The soils are generally shallow and gravelly. Area coverage is only .06 km^2.

Soils - The dominant soil is loamy skeletal lithic argiustolls in association with loamy skeletal lithic ustochrepts.

LAND CAPABILITY CLASSIFICATION AND LANDUSE PLANNING

The land of the sampled area of the Lower Alaknanda Basin was classified on the basis of its capability as assessed by the standard scheme of shavg who applied in 1920 in Malaya. The said scheme was slightly modified according to the type and mode of the terrain as given below:

S.No.	Land capability classes	Description
1.	Class I I_{cl} $(\overline{100})$	Land well suited to agriculture without any special practices, well protected and high potential for Sericitization and Chloritization. It is irrigated and terraced land. The land evaluation ratio is 108 (highest in the Alaknanda Basin) occurs under the category of 100 to formulate the -eva. factor i.e. I_{cl} $(\overline{>100})$
2.	Class II $II_{cl\ 2}$ $(\overline{80-100})$	Land suited to agriculture. Unirrigated agricultural land moderately potential for agricultural development with a restricted

		range of crops needing intensive management. The land evaluation ratio is 80 which occurs under the category of (80-100) to formulate -eva factor i.e. $II_{\frac{cl.2}{(80-100)}}$
3.	Class III $III/_{\frac{c2.3}{(60-80)}}$	Land is potential for forest development rationally suited to commercial timber exploitation. Land evaluation ratio is 78 which occurs under the category of (60-80) formulate the -eva factor i.e. $III_{\frac{c2.3}{(60-80)}}$
4.	Class IV $IV/_{\frac{c3.4}{(50-60)}}$	Land suited to forest conservation with little mineral agricultural or forest development potential. Land evaluation ratio is 54, which occurs under the range of 50-60 to formulate -eva factor, i.e. $IV_{\frac{c3.4}{(50-60)}}$
5.	Class V $V_{\frac{c3.4}{(40-50)}}$	Land needing an ecological tilt in its uses, highly suited to orchards and gardens. Land evaluation ratio is 50, which occurs under the category of (40-50) to formulate -eva factor i.e. $V_{\frac{c3.4}{(40-50)}}$
6.	Class VI $VI_{\frac{c3.4}{(<40)}}$	Land highly suitable for managed grass land, pastures. Its land evaluation ratio is 34 which occurs under category of 40 to formulate -eva factor i.e. $VI_{\frac{c3.4}{(>40)}}$
7.	Class VII $VII_{cl,2,3,4}$	Land uder this class consists of settlements, roads, ponds and other cultural uses notified as Miscellaneous land.

The said land capability classes possess highest -eva. ratio on the Alaknanda catchment area due to favourable soil genesis factors and land system units as characterised in the Tables

VIII and IX.

Soil Mapping and Landuse Survey for Sampled Villages in the Lower Alaknanda Basin

In the present study the remote sensing techniques were applied for soil mapping and landuse survey on large scale topographical maps land sat imagiries and photographs of the region were used for interpretation and delimitation of land system units, vegetation drainage and salient structures like thrusts and faults causitive to soil genesis. The soils may not be interpreted from the aerial photographs but their associates of physiographic elements like relief, slope, tone, texture, aspect and prominent lineaments (thrust and fault). The prominent physiographic sub-units of the Lower Alaknanda Basin consists of very Low Mountains, Low Mountains, River Valley Structures: (i) structural terraces and (ii) alluvial terraces, and North Almora thrust and Alaknanda thrust zones.

Case study of sampled villages

The villages of Gauchar, Bamoth and Augustmuni representing the Rudra-Karan Pedo Sericitic Zone of the Lower Alaknanda Basin were delineated from large scale maps and photographs to analyse soils and existing land use in order to develop landuse planning. The soil types of the villages as marked on the ground were recorded as 1-Hyper Thirmic Udic Haplustalls, 2. Hyper Thermic Typic Mollic Hypludalfs and Hyper Thermic Typic lender thants. The thick soils of the structural and alluvial terraces of the river valleys were developed by the commulative effect of mineral transformation caused by sericitization, chloritization, kaolinization and disintegration of rocks (as estimulated by the tectonic movements of the Alaknanda thrust and Almora thrust).

The region under investigation consisting of mountain ranges, trending from west to east have generally forested land on northern aspect and agricultural land on southern aspect on account of the degree of availability of the sunshine. It is worked out that the existing landuse of selected villages not only follow the nature of lithology and climate but it is also governed by the excess of human interference which creates ecological embalances through land depletion factors like landslides rock-slips, soil erosion and gulley formations.

After analysing the nature and type of land system of the region it is investigated that the Alaknanda and its tributary Mandakini have formed various broad river bands marked at

Gauchar, Bamoth, Nagrasu, Maletha and Augustmuni villages where alluvial terraces have been developed. There have been marked both structural and alluvial terraces in the said localities. Under river bond geomorphic conditions generally river forms the surfaces of alluvial deposits beside the slip of slope and escarpments (bluggs) on the opposite side where under cutting action of river current takes place.

To make a detailed study of the alluvial and structural terraces of the selected villages, soil samples were collected to work out their relationship with physiographic landuse as recorded below :

Village: Gauchar: $\frac{\text{Yon}}{6211}$ (Lower Alluvial Terrace)

Location	- Long. - Lat. 30'° 17' 19"/79'° 9' 18"
Land system unit	- Lower alluvial terrace; elevation 800 m.
Aspect	- North East.
Slope	- Flat to gentle
Vegetation	- Open-bushes and scrubs.
Drainage	- Well drained
Parental rock	- Alluvial deposits.
Soil	- Heavy loam (Typic Hapludalfs).
Landuse	- Cultivated (cropped).

Land capability class

1 Typical profile - $\frac{\text{Yon}}{6211}$

Morphological description

Depth (cm)	Description
0-12	Brown to dark brown (10YR 5/4) moist loam.
12-36	Yellowish brown (10YR 5/4) moist loam.
36-80	Yellowish brown (10YR 5/6) moist heavy loam.
80 +	Bed rock.

Village Gauchar: $\frac{\text{Hyr}}{6111}$ (Upper structural Terrace)

Location	- Long. Lat. 30'° 17' 19"/79'° 9' 18"
Land system unit	- Upper terrace, elevation 830 m.
Aspect	- North-east.
Slope	- Flat to gentle
Drainage	- Well drained.
Parental rock	- Alluvial deposits.
Soil	- Fine loam.
Landuse	- Cultivated

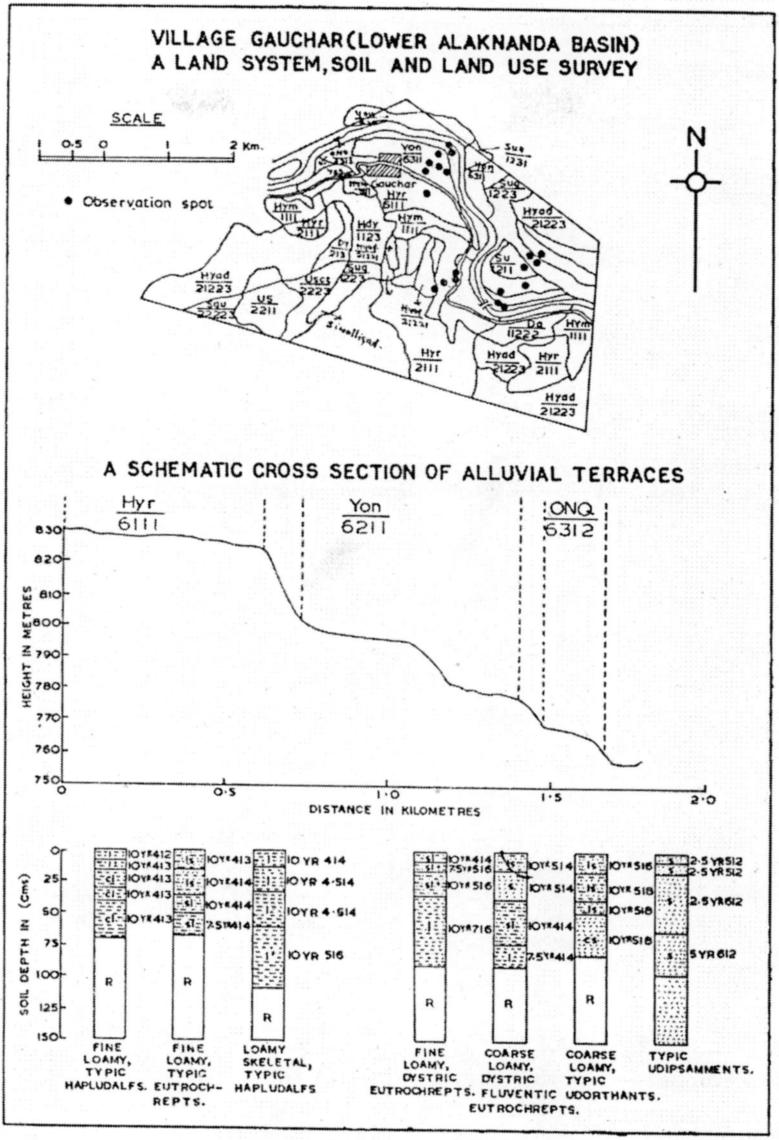

Fig. 8: Village Gaucher (Lower Alaknanda Basin) Land System, Soil and Land Use Survey

Land capability class

I Typical profile - $\frac{\text{Hyr}}{6111}$

Morphological description

Depth (cm)	Description
0-10	Dark greyish brown (10YR 4/2) moist laom.
10-18	Dark brown (10YR 4/3) moist heavy loam.
18-30	Dark brown (10YR 4/3) moist clay loam.
30-40	Dark brown (10YR 4/3) moist clay loam.
40-73	Dark brown (10YR 4/3) moist clay loam.
73+	Bed rock.

Village Bamoth: $\frac{\text{Von}}{6211}$ (Lower alluvial terrace, Low mountain)

Location	- Long. Lat. 79'^{o} 2' 30"/29'^{o} 23' 30"
Land system unit	- Very Lower alluvial terrace, elevation 780 m to 800m.
Aspect	- 55% with southern aspect.
Slope	- Flat to gentle slope - 0-80.
Vegetation	- Bushes and open scrub land with Amla trees.
Drainage	- Well drained.
Parental rock	- Alluvial deposition.
Soils	- Loamy sand (Typic Ustorthents).
Landuse	- Cultivated (cropped).
Land capability class	

GENERAL DESCRIPTION

The profile was studied in freshly dug pit on the south facing lower terrace of alluvial deposits. Generally the land which is not cultivated is clad with scrubs and bushes. The profile is being explained below:

I Typical profile No. $\frac{\text{Von}}{6211}$ Loamy sand (Typic Ustrothents)

Morphological description

Depth (cm)	Description
0-25	- Dark yellowish (10YR 4/4), moist loamy da sand.
25-38	- Dark yellowish (10YR 4/4) moist sandy loam, coarse.

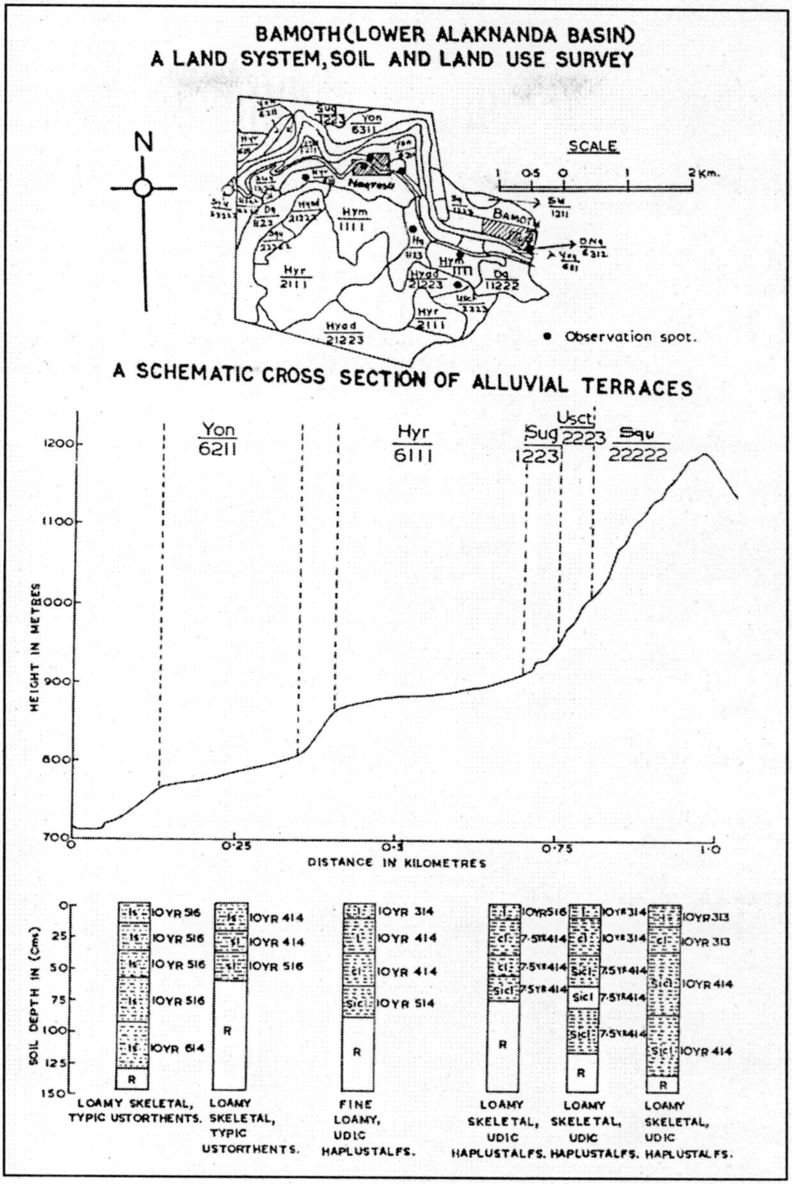

Fig. 9: Village Bamoth (Lower Alaknanda Basin) Land System, Soil and Land Use Survey

38-63 - Yellowish brown (10YR 5/6) moist sandy loam, coarse fragments more than 35%.
63+ - Bed rock.

II Typical profile No. $\frac{\text{Hyr}}{6111}$ - Fine loamy (Udic Hoplustalfs).

Morphological description

Depth (cm)	Description
0-12	Dark yellowish brown (10YR 3/4) moist loamy.
12-36	Dark yellowish brown (10YR 4/4) moist heavy loam.
36-62	Dark yellowish brown (10YR 4/4) moist clay loam.
62-88	Yellowish brown (10YR 5/4) moist silt clay loam.
88+	Bed

Village Bamoth: $\frac{\text{Squ}}{22222}$ (Low mountain)

Location	- Long. Lat. 79'° 2' 30"E/29'° 23' 30"
Land system unit	- Concave slope of low mountain.
Aspect	- 55% southern
Slope	- Moderate to steep.
Vegetation	- Medium forest
Drainage	- Well drained
Parental rock	- Phyllite and quartzite.
Soil	- Loamy skeletal (Haplustolls).
Landuse	- Forest

Land capability class

I Typical profile - $\frac{\text{Squ}}{22222}$ - Loamy skeletal (Haplustolls).

Morphological description

Depth (cm)	Description
0-18	Dark brown (10YR 3/3) moist loamy.
18-38	Dark brown (10YR 3/3) moist clayey loam.
38-88	Dark brown (10YR 4/4) moist silt clayey loam.
88-138	Dark yellowish brown (10YR 4/4) moist silt clayey loam.
138+	Bed rock

II Typical profile - $\frac{\text{Usct}}{2223}$ Loamy skeletal (Udic Haplustalfs).

Morphological description of profile

Depth (cm)	Description
0-8	Yellowish brown (10YR 5/6) Moist loamy.
8-33	Dark brown (7.4YR 4/4) moist clayey loam.
33-58	Dark brown (7.5YR 4/4) moist silt clayey loam.
58-75	Dark brown (7.5YR 4/4) moist silt clayey loam.
75-110	Dark brown (7.5YR 4/4) moist silt clayey loam.
110+	Bed rock.

Village Gauchar: $\frac{\text{ONO}}{6312}$

Location	- Long. Lat. 30'° 17' 19"/79'° 9' 18"
Land system unit	- (Recent alluvial flodd deposits), elevation 700 m
Aspect	- North North-east
Slope	- 8 per cent.
Vegetation	- Dvtubd (bushes with grass covers).
Drainage	- well drained.
Parent material	- Alluvial deposition.
Soil	- Coarse loamy (mixed Hyper thermic Typic Udorthents).
Landuse	- Cultivated cropped

Land capability class

GENERAL DESCRIPTION

The profile was studied in a freshly dug pit. These soils are found in the stream beds and restricted upto terraces of the same streams.

I Typical profile - $\frac{\text{ONO}}{6312}$

Morphological description

Horizon	Depth	Description
A_1	0-16	Yellowish brown (10YR 5/6) moist, loamy sand, structureless (single grain); non-sticky and nonplastic, loose, many, very fine roots; clear wavey boundary.
A_2	16-36	Yellowish brown (10YR 5/8) moist, loamy sand; structureless; non-sticky and non-

		plastic, loose; few, fine to medium.
A_3	36-48	Yellowish brown (10YR 5/8) moist, loamy sand; structureless; non-sticky and non-plastic, loose; common, medium roots; clear wavey boundary.
C	48-80	Yellowsih brown (10YR 5/8) moist, coarse sand; structureless; non-sticky and non-plastic, loose; few, medium roots.

RANGE OF CHARACTERISTICS

1. **Profile characteristics:** The depth of the solum varies from 60 cm to 100 cm. The depth of epipedon ranges from 10 cm to 20 cm. The coloured epipedon varies from dark brown to dark red. (10YR 3/3 to 2.5Y 3/2) and the exture varies from loamy sand to sand.

The colour of sub-surface horizon varies from dark yellowish brown to light grey and dominant texture is loamy and varies from loamy sand to coarse sand.

Landuse and management practices

Coarse grasses and thorny scrubs grown on the soils and are used as grazing lands.

Soil families

Fine loamy, coarse loamy sandy skeletal, loamy skeletal.

Associated soils

Typic Udic psamments, Typic psammaquents.

Gauchar village

The schematic cross section of the village Gauchar depicts a real shape of the alluvial terraces on the semi-circular bend of the river Alaknanda distinctly. There have been marked three distinct surfaces of alluvial deposits (Figs. 9) denoted by fraction $\frac{Hyr}{6111}$, $\frac{Yon}{6211}$, $\frac{ONO}{6312}$. On the other side of the river, digit number indicates a very conspicuous location of the terrace denoted by digit 6112. The numerater of the fraction as allotted for every land system sub-unit to evaluate its soils and landuse; indicates type, subgroup and family of the soils i.e. Hyr, Yon,

ONQ. Mixed Hyperthermic Typic Udorthents Typic Hapludolls and Typic Hapludalfs loamy skeletal fine loamy and coarse sandy. The upper alluvial terrace as denoted by $\frac{\text{Hyr}}{6111}$ depicts that it consists of maximum horizons with minimum thickness of 75 cm. fine loamy soils and its lowest terrace denoted by $\frac{\text{ONQ}}{6312}$ consists of four horizons with maximum thickness of 150 cm of coarse sandy loam (Fig.-7). The figures of denominator of the fraction indicate land system sub-unit aspect, slope landuse and its sub-categories. The first figure of the denominator of the fraction $\frac{\text{Hyr}}{6111}$ indicates the land system sub-unit i.e. river valley terrace, the second figure slope, third figure landuse and last figure landuse sub-category. The aspect has not been considered in the valley terrace.

The village Bamoth exists on the next annular bend of the river Alaknanda near Gauchar. The schematic cross section of the village indicating its terrace formation on the right bank of the river, depicts clearly two surfaces of the alluvial deposits as denoted by $\frac{\text{Yon}}{6211}$ and $\frac{\text{Hyr}}{6111}$. The denominators of the fractions, i.e. $\frac{\text{Suq}}{1223}$, $\frac{\text{US}}{2223}$, and $\frac{\text{Squ}}{22222}$ indicate moderate to steep slope of very low and low mountains ranging from 1000 to 2000 meter high from m.s.l. The soil profiles of the corresponding landsystem sub-unit indicate that the maximum depth of soil i.e. 125 cm occurs under the fraction $\frac{\text{Squ}}{22222}$ because it is structural terrace clad with dense forest where soil is formed by both mechanical and chemical processes of sericitization, chloritization (mineral transformation) and rock disintegration (estimulated by the North Almora thrust). The thickness of soils on the second terrace (50 cm) as denoted by the fraction $\frac{\text{Yon}}{6211}$ is less than that first terrace (80 cm) because of its severe soil erosion. The corresponding terrace bears moderate to steep slope which causes severe less of soil and water with the result the last surface of the alluvial deposit near the river consists of maximum deposits bearing 125 cm thickness of loamy skeletal soils (Fig. 8).

Both the terraces of the village have flat to gentle and moderate slope with agricultural lands. But soils of mountain clad with dense forests are deep i.e. 130 cm because it is originally formed under mineral transformation processes and well protected by the thick vegetal cover.

Village Augustmuni: $\frac{\text{Von}}{6211}$

Location - Long. Lat. 79" 2'E/29'° 23' 45"N .
Land system unit - Lower Alluvial terrace, elevation 750 m.
Aspect - North-east.
Slope - Flat to gentle 0°-80°.
Vegetation - Open scrub land and oak trees.
Drainage - Well drained.
Parental rock - Alluvial deposits.
Soil - Fine loamy (Dystric Fluventic Eutrochrepts).
Landuse - Cultivated (cropped).

Land capability class

GENERAL DESCRIPTION

Profile was selected in the alluvial deposits under moist conditions. The soils of the lower terraces consists of flat to gentle slope facing north-east. Soils are moderately eroded due to grazing over lapping and torrential runoff of the seasonal streams.

I Typical profile - $\frac{\text{Von}}{6211}$ Fine loamy (Fluventic Eutrochrepts).

Morphological description

Depth (cm)	Description
0-12	Light olive brown (2.5YR 5/4) moist sandy loam.
12-25	Greyish brown (10YR 5/2) moist heavy loam.
25-63	Yellowish brown (10YR 5/6) moist loam.
63-100	Light olive brown (2.57 5/4) moist loamy sand
100+	Bed rock.

Village Augustmuni: $\frac{\text{Hym}}{1111}$

Location - Long. Lat. 79'° 2'E/29'° 23' 43"N
Land system unit - upper structural terrace, elevation 850 to 950 m.
Aspect - North-east
Slope - Gentle to moderate
Vegetation - Open scrub land with broad-leaved forest.
Drainage - Well drained
Parental rock - Quartzite/phyllite.
Soil - Fine loamy (Typic argiudolls)

More chemical weathering (mineral transformation sericitization) and less mechanical.

Landuse Cultivated (fallow).

Land capability class

GENERAL DESCRIPTION

The profile was selected from the structural terrace with gentle to moderate slope with elevation ranging from 850 to 900 m. The terrace is composed of original soils chemically transformed from quartzite and phyllite parental rocks by mineral transformation process - sericitization and rock disintegration. (Fig. 9).

I Typic profile - $\frac{\text{Hym}}{\text{IIII}}$ - Fine loamy (Typic Argiudolls)

Morphological description

Depth(cm)	Description
0-28	Very dark, greyish brown (10 YR 3/2) moist loam.
28-62	Dark brown (7.5YR 4/4) moist clayey loam.
62-94	Brown (7.5YR 5/4) moist heavy clayey loam.
94-120	Brown (7.5YR 5/4) moist clayey loam.
120+	Bed rock.
	Village Augustmuni : $\frac{\text{Hyad}}{21222}$

Location - Long.lat 79'° 2'E/29'° 23'45"
Land system unit - Low mountain, elevation 1200 m.
Aspect - Northern
Slope - Flat to gentle
Vegetation - Medium forest.
Drainage - Drained but slight erosion occurs.
Parental rock - Quartzite and phyllite.
Soil genetic process - More pedo-chemical, mineral transformation (serticitization) and less mechanical.
Soil - Fine loamy (Typic Hapludalf).
Landuse - Forest - medium.
Land capability class

GENERAL DESCRIPTION

The profile was selected on the low mountain to study the soil genetic processes and their relationship with physiography and landuse. Pedo-chemical and mineral transformation chemical

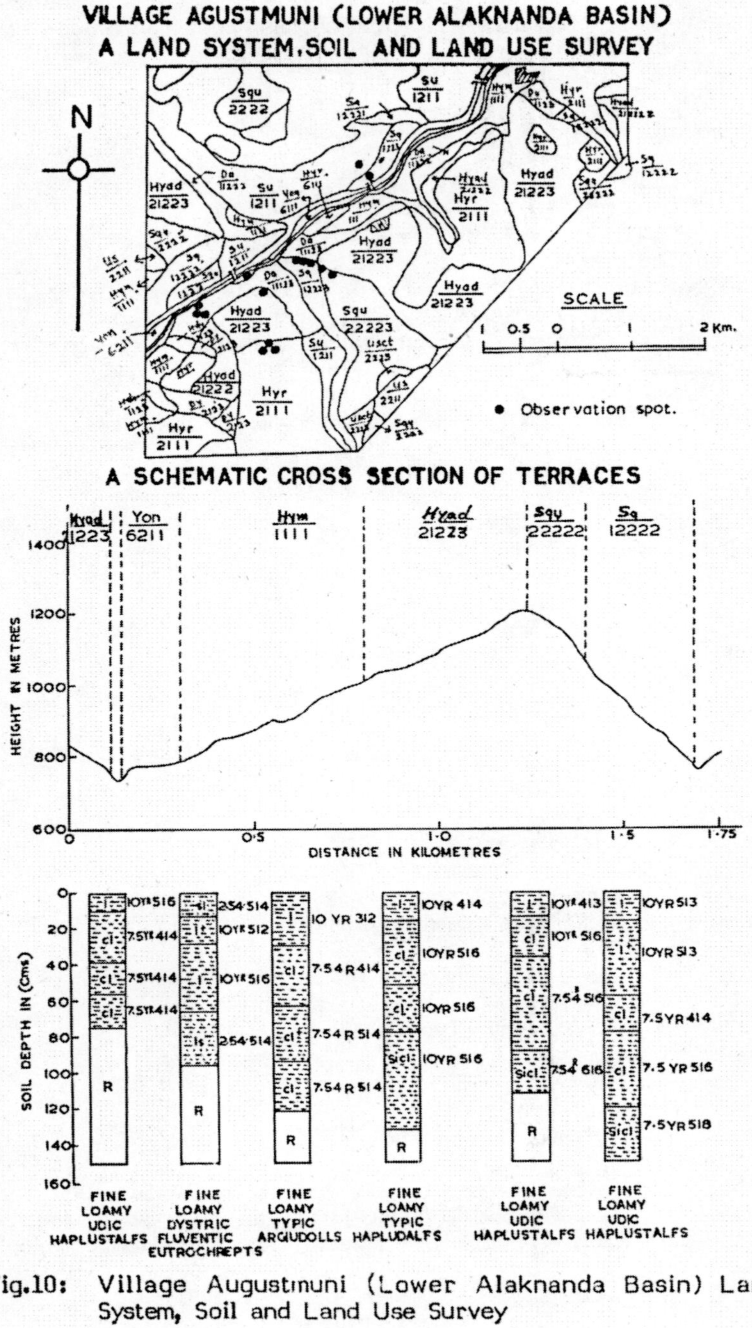

Fig.10: Village Augustmuni (Lower Alaknanda Basin) Land System, Soil and Land Use Survey

process (sericitization) and less mechanical weathering occur throughout the year with the result the depth of the original soil is found more than alluvial deposits.

I Typic profile - $\frac{\text{Hyad}}{21222}$

Morphological description

Depth(cm)	Description
0-17	Dark yellowish brown (10YR 4/4) moist loam.
17-80	Yellowish brown (10YR 5/6) moist.
30-78	Yellowish brown (10YR 5/6) moist.
78-140	Yellowish brown (10YR 5/6) moist silty clay loam.
140+	Bed rock.

The village Augustmuni lying on the left bank of the river Mandakini comprises of broad terraces in the lower Alaknanda Basin. The schematic cross section of the village (Fig. 10) indicates that the lowest terrace denoted by the fraction of $\frac{\text{Yon}}{6211}$ is just adjacent to the river channel and it is broader than that of the upper structural terrace denoted by the fraction of $\frac{\text{Hym}}{1111}$. The numerator of the fraction of $\frac{\text{Yon}}{6211}$ depicts fine loamy soil (Dystric Fluventic Eutrochrept) characterising its depth, colour and texture i.e. 100 cm light olive brown 2.5Y 5/4 to yellowish brown 10YR 5/6 moist with loam and loamy sand respectively. The first figure of the denomerator of the fraction $\frac{\text{Yon}}{6111}$ itself indicates land system sub-units i.e. valley structure, the second figure indicates alluvial terrace and the third slope (ranging flat to gentle) and last figure depicts the landuse (irrigated agricultural land).

Similarly the upper structural terrace denoted by the fraction $\frac{\text{Hym}}{1111}$ bearing fine loamy and loamy skeletal (Typic Hapludalfs, Dystric Eutrochrepts, Mollic Hapludalfs) is characterized with moderate to strongly sloping under unirrigated agricultural single crop land adjacent to the low mountains clad with medium forest land system.

The study of soil horizons of each sub-units denoted by fractions $\frac{\text{Dy}}{2123}$, $\frac{\text{Squ}}{22222}$ and $\frac{\text{sq}}{12222}$ indicate that their soils are fine loamy (Typic Hapludalfs), loamy skeletal (Udic Haplustalfs

Udic Argiustolls) and fine loamy (Udic Haplustalfs, Udic Argiustolls) which have been well developed on the low mountains (Quartzites and phyllites) clad with dense and medium soils. Soils are very deep. The soil profile of the fraction $\frac{\text{Squ}}{2222}$ consists of five horizons with maximum thickness of soil i.e. 140 cm.

After analysing the schematic cross sections of the said land system sub-units, it is worked out that the depth of the soils tends to increase from alluvial terrace to structural terrace due to the mineral transformation process of its soil genesis. By sericitization process quartzites is usually transformed into sericitic which finally comes into soils. Thus the thickness of soils now depends on the account of quartzite and the degree of sericitization and soil protection. Both the alluvial and structural terraces of the villages facing southward are generally occupied by agricultural land. The low and very low mountains facing northwards are clad with dense forests.

The moderate to steep sloping mountain spurs, ridges, talus cones facing southward are generally found under scrubs and grasses. The newly formed repose slopes of the talus cones are found depleted in the whole valley. They may be protected from overgrazing and overfallowing by introducing certain measures recommended in the text.

APPENDIX I

Evaluation of Soil and Land Systems for Landuse Planning

S. No.	Mapping Units SOI/Land Systems	Land Evaluation Factor	Area Km^2	Percent-age to	Existing Land Use	Recommended Use
1	Hym/111	180	21.5625	3.10	Cultivated Land	Cultivated
2	Da/112	185	1006	1.45	Forest	Forest
3	Hdy/113	175	4.68	.67	Scrubs/grass land	Grass Lands/ Pasture development
4	Su/121	185	18.37	2.64	Cultivated	Cultivated
5	Sg/122	160	8.12	1.17	Forest	Forest
6	Sug/123	160	13.12	1.89	Scrubs	Forest
7	Hyr/211	195	119.00	17.09	Cultivated	Cultivated
8	Hyad/212	190	120.25	17.28	Forest	Forest
9	Dy/213	180	18.56	2.68	Scrubs	Pasture
10	glls/221	155.25	78.18	12.22	Forest	Forest
11	Sgu/222	134.75	85.06	11.23	Cultivated	Cultivated/ Plantation
12	Usct/223	134.75	26.56	3.82	Scrubs	Forest
13	Tc/224	130	.06	0.008	Barren	Grass Land
14	Ha/331	165	2.26	0.38	Cultivated	Cultivated
15	Ahd/312	165	82.18	11.81	Forest	Forest

S. No.	Mapping Units SOI/Land Systems	Land Evaluation Factor	Area Km^2	Percent-age to	Existing Land Use	Recommended Use
16	Ha/313	165	4.37	0.63	Scrubs/ grassland	Cultivated
17	Lwt/314	125	.62	0.09	Barren	Grass land/ Plantation
18	Si/321	120	4.12	0.59	Cultivated	Pastures/ Plantation
19	Sui/322	152	45.37	6.52	Forest	Forest
20	Set/323	120	4.93	0.71	Scrubs	Grassland/ development
21	Te/324	120	1.50	0.22	Barren	Grasslands/ pastures
22	Dak/412	155	1.31	0.19	Forest	Forest
23	Kd/413	150	0.25	0.04	Scrubs	Forest
24	W/414	150	1.87	0.27	Barren	Grass lands/ plantation
25	Ks/422	125	.56	0.08	Forest	Forest
26	Tew/424	115	1.68	0.24	Barren	Grasslands
27	Hyr/531	260	5.43	.78	Cultivated	Cultivated
28	Yon/541	255	7.37	1.06	Cultivated	Cultivated
29	Ong/55	255	.87	0.13	Cultivated	Cultivated
30	Escarpments	104	6.50	0.93	Scrubs	Forest
	TOTAL	696.00	100%			

APPENDIX II

Terrain Evaluation for Landuse Planning in the case study area of Garhwal Dudhaloli/group in Lower Alaknanda Basin

S. No.	Mapping Units Soil/Land Systems	Land use	Area Km^2	Percent-age	Land Evaluation Factor	Existing Landuse	Recommended Landuse
1	Hym/11211	Very Low Mt Northern Aspect cultivated	21.56	3.10	180	Cultivated	Cultivated
2.	Da/11221	Sparse Forest	1.87	0.27	185	Forest	Forest
3.	Da/11222	Medium Forest	5.56	0.80	185	Forest	Forest
4.	Da/11123	Dense Forest	2.62	0.38	185	Forest	Forest
5.	Hdy/1123	Open Scrub/ grasslands	4.68	0.67	175	Open Scrub	Grasslands/ Pasture development
	12	Southern Aspect					
6.	Su/11211	Cultivated	18.37	2.64	185	Cult	Cult
7.	Sq/12121	Sparse Forest	3.31	0.47	160	Sparse Forest	Medium Forest
8.	Sq/12222	Medium Forest	4.00	0.57	160	Medium Forest	Dense Forest
9.	Sq/12223	Dense Forest	0.81	0.12	160	Dense Forest	Dense Forest
10.	Sug/1213	Open Scrub/ grassland	13.12	1.89	160	Scrubs	Sparse Forest
	2 21	Low Mountain Northern Aspect					

S. No.	Mapping Units Soil/Land Systems	Land use	Area Km²	Percent-age	Land Evaluation Factor	Existing Landuse	Recommended Landuse
11.	Hyr/2111	Cultivated	119.00	17.10	195	Cultivated	Cultivated
12.	Hyad/21221	Sparse Forest	4.18	0.60	190	Sparse	Medium Forest
13	Hyad21222	Medium Forest	22.37	3.21	190	Medium Forest	Dense Forest
14	Hyad/21223	Dense Forest	93.68	13.46	190	Dense Forest	Dense Forest
15	Dy/2123	Open scrubs/ Grassland	18.56	2.67	180	Scrubs/ Grassland	Grassland/ pasture development
	22	Southern Aspect					
16.	Us/2211	Cultivated	78.18	11.23	155.25	Cultivated	Cultivated
17	Sgu/22221	Sparse Forest	8.12	1.17	155.25	Sparse Forest	Medium Forest
18	Sgu/22222	Medium Forest	56.12	8.06	134.75	Medium Forest	Dense Forest
19	Sgu/22223	Dense Forest	20.81	2.99	134.75	Dense Forest	Dense Forest
20	Uset/2223	Open scrub/ grassland	26.56	3.82	134.75	Open scrubs/ grassland	Sparse Forest grassland
21	Uset/22344	Barren/grass	0.05	0.01	130.00	Barren	
	3	High Mountains					
	31	Northern Aspect					
22	Ha/3121	Cultivated	2.62	0.38	165.00	Cultivated	Cultivated
23	Ahd/31221	Sparse Forest	1.62	0.23	165.00	Sparse Forest	Dense Forest

S. No.	Mapping Units Soil/Land Systems	Land use	Area Km^2	Percent-age	Land Evaluation Factor	Existing Landuse	Recommended Landuse
24	Ahd/31222	Medium forest	6.37	0.92	165.00	Medium Forest	Dense Forest
25	Ahd/31223	Dense Forest	74.18	10.66	165.00	Dense Forest	Dense Forest
26	Ha/3123	Open scrub/ grassland	4.37	0.63	165.00	Open scrubs grasslands	Cultivated
27	Lwt/31345	Barren rock out more than 35% of the area	0.62	0.09	125.00	Barren	Grassland/ plantation
	32	<u>Southern Aspect</u>					
28	Sl/3211	Cultivated	4.12	0.59	120.00	Cultivated	Cultivated
29.	Sui/32221	Sparse Forest	3.56	0.51	152.00	Sparse Forest	Medium Forest
30	Sui/32122	Medium Forest	22.06	3.17	152.00	Medium Forest	Dense Forest
31	Sui/32223	Dense Forest	19.75	2.84	152.00	Dense Forest	Dense Forest
32	Set/3223	Open scrub/ grassland	4.93	0.71	120.00	Open scrub/ grassland	Grassland/ pasture development
33	Tc/32344	Barren mixed with grass	1.31	0.18	120.00	Barren	Grassland/ Pasture
34	Tc/32345	Barren rock out crops more than 33% of the area	0.18	0.03	120.00	Barren	Grassland/ pasture
	4 41	<u>Very High Mt.</u> <u>Northern Aspect</u>					

S. No.	Mapping Units Soil/Land Systems	Land use	Area Km^2	Percent-age	Land Evaluation Factor	Existing Landuse	Recommended Landuse
35	Dak/41222	Medium Forest	0.12	0.02	155.00	Medium Forest	Dense Forest
36	Dak/41223	Dense Forest	1.18	0.17	155.00	Dense Forest	Dense Forest
37.	Kd/4133	Open scrub/ Land	0.25	0.04	150.00	Scrubs	Grassland/ sparse Forest
38.	W/41344	Barren mixed with grasses	0.81	0.12	150.00	Barren	Grassland/ sparse Forest
39.	W/41345	Barren rockout crop nore than 33% of the area	1.06	0.15	150.00	Barren	Grassland/ sparse Forest
40	12 Ks/42322	Southern Aspect / Medium Forest	0.56	0.08	125	Medium Forest	Dense Forest
41	Tew/42345	Barren rock outcrops 33%	1.68	0.24	115	Barren	Grassland
42	5 513 Hyr/5131	Upper Terrace / Upper Terrace Cultivated	5.43	0.78	260	Cultivated	Cultivated
43	514 Yon/5141	Lower Terrace Cultivated	7.37	1.06	255	Cultivated	Cultivated
44	ONQ/515	Flood plain	0.87	0.13	255	Cultivated	Cultivated
45	Escarpment	Escarpment	0.75	0.11	104		Scrubs/grass land
46	River bed	6.50	0.93				
		696.00	100%				

BIBLIOGRAPHY

Acharya, K.C.S.,1983: **Food Security System of India: Evolution of Buffer Stocking Policy and Evaluation,** New Delhi: Concept Publishing Company.

Atkinson, E.T.,1882: **The Himalayan Distt. of N.W. of India,** 3, Vols. Govt. Press, Allahabad, 84.

Auden, J.B. 1937 : **The Structure of Himalaya in Calcutta.**

Basu, T.K. 1942 : **Soil fertility.**

Beckett, S. & D.,1886 : Report on the Settlement Operations in Garhwal Distt. Govt. Press, Allahabad.

Berry, B.J.L., 1978 : **Perspectives in Geography :** The Nature **change in Geographical Ideas.**

Bhalla, G.S. and G.K. Chadha, 1983 : **Green Revolution and the Small Peasant : A Study of Income Distribution among Punjab Cultivators,** New Delhi : Concept Publishing Company.

Bhatia, S.S. 1967 : A New Measure of Agricultural Efficiency in U.P. in **India Economic Geography,** Vol. 43, No.3, p. 248.

Bhattacharjee,J.C. 1982 : "Bioclimate Analysis of India" **Bull. No.7** Ed. R.S. Murthy. N.B. S.S. New UK Planning Nagpur.

Briggs, David, 1985: Agriculture and Environment the Physical Geog. of Temp. Agri. System.

Buchmen & Bredy, 1961 Nature & Prospects of soils.

Carson, M.A. 1972 : **Hill Slope Form and Process.** Cambridge University Press.

Charley, R.J. 1959 : The Geomorphic significance of some Oxford Soils. **A.M.L, Sci.** 252.

Carlstein, T., 1982 : Time Resources, Society and Ecology.

Dabral, S.P. **Alaknanda Upattika,** Dogadda, Virgatha Prakashan.

Dabral, S.P. "Uttarakhand ke Bhotantik Garhwal" Dogadda Garhwal.

Dabral, S.P. **Uttarakhand Ke Pashucharak Dogadda Garhwal.**

Dabral, S.P. **Uttarakhand Ka Itihas,** Dogadda.Garhwal Pt.I & II.
Dabral, S.P. 1968: Uttarakhand Ka Itihas, Dogadda Garhwal Pt.III & IV.

Daly, M.T. 1972 : Techniques and Concepts in Geography.

Dewan, M.L., 1982 : **Agriculture and Development in India : A Case Study on Dignity of Labour,** New Delhi : Concept Publishing Co.

Enyedi, G.Y. Geographical Type of Agriculture, Applied Geography in Hungary Budapress, 1964.

Everson, J.A., 1969 : **Concept in Geography** Settlement Pattern.

Gatrell, A., 1984 : Distance and space : A Geographical Perspective.

Gupta, I.C. 1983 : **Twentieth Century of Soil Salinity Research**

in India : An Annolated Bibliography, New Delhi : Concept Pub. Co.

Gupta, R.K., 1981 : Ecological Approach to Land Capability Classification and Regional Planning **: Perspectives in Agricultural Geography,** Vol. 3. 89-108 Ed: Noor Mohammad. Concept Publishing Company, New Delhi.

Gupta, R.K., 1981 : Water shed Management Practices in the Himalayas for Rural Development in Mts. ed. J.S. Singh Nainital.

Gupta, S.K., 1983 : **Global Research on Drainage in Agriculture,** New Delhi: Concept Pub. Co. (in Press)

Gupta,R.K. 1983 : Reconstruction of Himalayan Ecosystem Through Horticulture for Economic-Development and Water shed Management, Himalayan Mt. & Men: Studies in Eco. Devl. Ed. T.V. Singh.

Hartly, CWS. 1949 : Soil Erosion in Malaya Corora, 1-2.

Har Prasad, 1984 : A Note on the Quest for Geographical Theory, N. **Geog. Journal of India,** Vol. XXX,4, Dec. pp. 231-35, Varanasi.

Har Prasad, 1972 : Integrated Area Development and its Appraisal in Dehradun, **Indian Journal of Regional Science,** Vol. IV.

Har Prasad, 1973 : Integrated Approach in the Synthesis of Natural Factors and Analysis of Land System in the Dehradun.

Har Prasad, 1977 : The Dehradun : A Methodological study in Regional Land use Ecology. **The Geographical Observor,** Vol. 13.

Har Prasad, 1978 : The Dehradun : A Micro-regional study with Special Reference to Integrated Area Development. **Abstracts Journal ICSSR,** Vol. V, No. 1 & 2.

Hayden, H.H., 1933 : Geography and Geology of Himalaya Mts. and Tibet. Govt.of India Press, Delhi.

Harriss, Barbara, 1984 : State and Market : **State Intervention in Agricultural Exchange in Dry Region of Tamil Nadu,** New Delhi, Concept Pub. Co.

Kostrowicks, J. 1940 : Geographical Typology of Agriculture Principles and Methods. **Geogr. Polonica.2.** p. 67.

Kostrowicks, J. Principles, Basic Nations and Criteria of Agricultural Typology.

Mandal, R.B., 1980 : **Recent Trends and Concepts in Geography** (3 eds.), New Delhi : Concept Publishing Co.

Mishra, Pratibha, 1984 : **Soil Productivity and Crop Potential : A case Study (Dist. Alwar, Rajasthan, New Delhi** : Concept Pub. Co.

Misra, R.P., **Rural Development : Capitalist and Socialist Paths** (5 Volumes), New Delhi : Concept Pub. Co.

Morgan, R.P.C. 1976 : **The Role of Climate in the Denunation**

System. A case study from West Malaysia, in Derbeshie E. John. Wiley, London, pp.317-43.

Morgan, R.P.C. 1979 : **Topics in Applied Geography Soil Erosion,** Longman, London.

Morgan, W.B. 1969 : The Zoning of landuse Around Rural Settlements in Tropical Affairs.

Munten, R.J.C. & Norris, J.M. **The Analsysis of Farm Organization. An Approach to the Classification of Agricultural Land in Britain.**

Murthy, R.S. 1982 : Remote Sensing Techniques in Establishing Geomorphic Soil Relationships, **Photo Nirvachak Jour.** vol. 10 No. 2. 1-11.

Murthy R.S. 1983 : Application of Remote Sensing Technique for Land Evaluation and Classification for Agriculture. **Proc. of Ind. Academy of Science.**

National Academy of Science Resource Sencing from Space Respects for Development Centres, 1968 : Washington, D.C., USA.

National Bureau of Soil Survey and Landuse Planning : Soil Taxonomy : A Basic System for Making and Interpreting Soil Surveys.

Nitya Nand, 1977 : **U.P. Himalaya - A study of Regional Geography edited** by R.L. Singh.

Nitya Nand, Crop Combination in Rajasthan, **Geographical Review of India,** Vol. XXXIV, No.1, pp. 46-60.

Pravana Nand, 1980: Explorations in Tibet, Culcutta Univ. Press.

Prohit, A.N. 1983: Garhwal Himalaya - A Quarrying Eco-System Studies in Eco-Development Himalayas In Himalayan Mt. and Men : Studies in Eco-Dev. Ed. T.V.Singh.

Saxena, P.B. 1978 : A Study of Land sliding and its land depletion in the Alaknanda valley (Garwal Himalaya) published in the **Himalayan Geology,** Vol. IX Nov.

Saxena, P.B. A Study of Environmental Resource Systems and Their Management Programme in Dehradun Valley - Key Note Address - presented in workshop for geomorphological & Environmental Development, Garhwal University Srinagar.

Saxena, P.B. 1977 : A Study of Drainage Morphometry for Resource Planning in the Alaknanda Valley (Garhwal himalaya) **Indian Journal of Regional Science,** Vol. IX, no. 2.

Saxena, P.B.1983: The Solani Ganga Interfluve : An Application of Remote Sensing Techniques on Soil Mapping for Agricultural Planning, Photo Niwachika Institute of Remote sensing, Dehra Dun, Vol. II, No. 1.

Saxena, P.B. 1980: A Geographical Study of Land Slides its Genesis Intensity of soil depletion and Zoning in the Alaknanda Valley (Garhwal Himalaya) published in **Perspectives in 'Geomorphology',** edited by Prof. H.S. Sharma, Rajasthan Univ., Jaipur, 1980.

Saxena, P.B. 1981: A Study of the Drainage Pattern in Dehradun Valley (Garhwal Himalaya) **Advanced Study in Geology,** No.12, Vol. Jan.19-21.

Saxena, P.B. 1970 : Population and food supply in Tehri Garhwal. **The Geographical Knowledge,** Vol.3, no.1-2, Jan-July.

Saxena, P.B. 1974 : Water Resources and Potential Utilization Programme in the Asan Valley Garhwal Himalaya. **The Geographical Observer,** Vol. 10, March.

Saxena, P.B. 1980: A Resource Appraisal of Springs of Krols in Dehradun Valley. **National Geographer Himalayan Geology** Vol.X.

Saxena. P.B. 1972: Trends of Spatial Distribution of Population Dehradun Valley, **The Geographical Knowledge,** Vol.V.

Saxena P.B. 1982: A Study of the Morphometric Determinants of the stage of Cycle of Erosion in the Nayar Basin, **Perspectives in Geomorphology,** edited H.S. Sharma Vol. IV.

Shafi, M., 1984 : **Agricultural Productivity and Regional Imbalances : A Study of Uttar Pradesh,** New Delhi : Concept Pub.Co.

Shafi, M. Measurement of Agricultural Productivity of the Great Indian Plains, **The Geographer,** Vol. XIX, No.1, pp. 4-13.

Sharma, Hari Shankar, 1979 : **The Physiography of the Lower Chambal Valley and its Agricultural Development : A Study in Applied Geomorphology,** New Delhi : Concept Pub.Co.

Sharma, P.R. Crop Cultivation Intensity, their Ranking and Crop Assocation Region in Chhatishgarh Region. A Geographical Analysis, **National Geog. Jour. of India,** Vol. XVIII, Pt. 2, pp. 91-101.

Sinha, B.N. 1968 : Agricultural Efficiency in India, **The Geographer,** Vol. XV, Nor. Special No. xxi International Geographical, Congress, pp. 101-127.

Singh Jasbir 1972 : A New Technique of Measuring Agricultural Productivity in Haryana (India). **The Geographer,** Vol. XIX, pp. 14-33.

Singh, B.B.: Graphing Pattern of Braut Block. A Temporal variation, **Geographical Observer,** Vol.9,pp. 51-60.

Strachy, J. 1953: On the Himalayas in Kumaon and Garhwal, Col. Rev. p. 120

Swaminathan, M.S. 1983 : **Science and Conquest of Hunger,** New Delhi : Concept Pub. Co.

Sphare, S.G. and Deshpande, Inter District Variation in Agricultural Efficiency in Maharashtra State: **Indian Journal of Agricultural Economics,** Vol. 19, No.1, pp. 242-252.

Trivedi, R.K. 1984: Practical Methods in Ecology and Environmental Science.

Vidyarthi, L.P. 1979: **Changing Patterns and Habits : A Socio-Cultural Study of Bihar,** New Delhi: Concept. Pub. Co.

Vink, A.P.A. 1958: Areial Photographs and the Soil Sciences in Aerial Surveys & Integrated Studies UNESCO, Paris, 127.31.

Vink, A.P.A. 1983 : **Land Scape Ecology and Land use** Ed. Davisian, D.A. Longman.

Williams, A.R. and Morgan, R.P.C. 1976 : Geomorphological Mapping Applied to Soil Erosion Evaluation, **J. Soil and Wat. Conserv.** 31, 1964-8.

Zaidi, Syed Sajid Husain, 1982: **Rural India and Malnutrition, Implications, Problems and Prospects,** New Delhi. Concept. Pub. Co.

Rovert, D. 1969: Rudal Remote Sensing a Better View Duxbury Press Massachussettes, USA.

Index